From Hiroshima to Fukushima to You

FROM HIROSHIMA TO FUKUSHIMA TO YOU

A PRIMER ON RADIATION AND HEALTH

Dr. Dale Dewar and Florian Oelck

for Physicians for Global Survival

Between the Lines
Toronto

First published in 2014 by
Between the Lines
401 Richmond Street West
Studio 277
Toronto, Ontario M5V 3A8
Canada
1-800-718-7201
www.btlbooks.com

Library and Archives Canada Cataloguing in Publication

Dewar, Dale, 1944–, author
    From Hiroshima to Fukushima to you : a primer on radiation and health / Dale Dewar and Florian Oelck.

Includes bibliographical references and index.
Issued in print and electronic formats.
ISBN 978-1-77113-127-8 (pbk.). — ISBN 978-1-77113-128-5 (epub).
ISBN 978-1-77113-129-2 (pdf)

1. Radiation — Health aspects.   I. Oelck, Florian, 1986–, author
II. Title.

RA1231.R2D49   2014   363.17'99   C2014-900287-4
C2014-900288-2

Cover design by Gordon Robertson
Printed in Canada

PHYSICIANS FOR GLOBAL SURVIVAL
MÉDECINS POUR LA SURVIE MONDIALE
CANADA

RECYCLED
Paper made from
recycled material
FSC® C103567

The research and writing of this book was made possible through generous support from the Physicians for Global Survival.

Between the Lines gratefully acknowledges assistance for its publishing activities from the Canada Council for the Arts, the Ontario Arts Council, the Government of Ontario through the Ontario Book Publishers Tax Credit program and through the Ontario Book Initiative, and the Government of Canada through the Canada Book Fund.

Canada Council
for the Arts
Conseil des Arts
du Canada

Canadä

ONTARIO ARTS COUNCIL
CONSEIL DES ARTS DE L'ONTARIO
50 YEARS OF ONTARIO GOVERNMENT SUPPORT OF THE ARTS
50 ANS DE SOUTIEN DU GOUVERNEMENT DE L'ONTARIO AUX ARTS

To the hundreds of thousands of people whose lives have been adversely affected by ionizing radiation: victims of atomic bomb testing and use, experimental medical use, and nuclear power plant accidents. And to the thousands who have benefitted through medical or industrial use.

— D.D.

For my mom, Dr. Elke Steinmann-Oelck, who lost her battle with breast cancer in 2012 while we were writing this book. "All that I am or ever hope to be, I owe to my angel mother." – Abraham Lincoln

— F.O.

Contents

CHAPTER 3: RADIATION AND THE HUMAN BODY

CHAPTER 4: RADIATION IN MEDICINE

CHAPTER 5: INDUSTRIAL USE OF RADIATION

CHAPTER 6: NUCLEAR POWER PLANTS

CHAPTER 7: URANIUM MINING

Preface

THIS BOOK was researched and written with the generous financial support of the Physicians for Global Survival (PGS), the Canadian affiliate of International Physicians for Prevention of Nuclear War (IPPNW). As such, it is cautionary with regard to ionizing radiation. It provides a forum for the historic voice of medicine, rooted in Hippocrates: "first and foremost, do no harm."

Education is crucial to understanding the effects of ionizing radiation on human health and the environment. PGS and its global affiliates through IPPNW are at the forefront of a campaign to provide information for both the general public and policy-makers.

The twentieth century has seen a military-industrial love affair with the power of the atom. By the 1970s, there were 67,000 nuclear bombs located largely in Russia and the United States. While most of these were pointed at one another, they were enough to destroy the planet thousands of times over. At that time, physicians became concerned. It was clear that in the event of a nuclear war, their clinical skills, which were so highly regarded in other emergency situations, would have no value. There could be no medical response.

Cardiologist Dr. Bernard Lown was the US inventor of an electrical defibrillator for the heart. He was also an early opponent of the nuclear arms race, who, along with Dr. Helen Caldicott, an Australian pediatrician who was then a professor at Harvard, founded Physicians for Social Responsibility. Dr. Lown found his counterpart in fellow cardiologist Dr. Yevgeniy Chazov, Deputy Minister of Health in Russia. In 1980, the two men met in Switzerland where they cofounded IPPNW. By the time the organization was awarded the Nobel Peace Prize in 1985, there were more than 135,000 members in forty-one countries.[1] They subsequently brought sufficient pressure to bear upon presidents Mikhail Gorbachev and Ronald Reagan for the two of them to meet in Iceland in 1986. Negotiations began that eventually led to the end of the arms race and a reduction in the total number of nuclear weapons in the world. Although destruction of the stockpile eventually stalled, IPPNW and its affiliates have continued to press for complete abolition.

For years, despite concerns about nuclear proliferation, IPPNW remained a cautious supporter of nuclear power. Following several catastrophic nuclear events, the organization reversed its opinion. On the heels of the accident at the Three Mile Island nuclear power plant near Harrisburg, Pennsylvania, in March 1979, the Chernobyl nuclear disaster in April 1986 affected a large population and several countries. Physicians were on the frontline during Chernobyl's aftermath. While they were caring for victims suffering from the effects of radiation, they gathered information and performed research. They hoped to present their findings at the 1996 International Atomic Energy Agency (IAEA) and United

Nations-sponsored "One Decade after Chernobyl" conference in Vienna. Instead, it was closed to the general public. In response, physicians and other concerned citizens organized a "shadow conference" at the same time, called "A Decade after Chernobyl, Summing up the Consequence of the Accident."[2]

The secrecy surrounding information and decision-making meetings and the undemocratic suppression of clinical information that contradicted the IAEA's position that only four thousand people had or would die as a result of the accident[3] led enough physicians, by 1998, to conclude that the political, health, and environmental risks of nuclear power were high enough that it should be phased out. That year, IPPNW tabled a motion opposing nuclear power. Since uranium has only two uses, nuclear power and nuclear weapons, the organization called for an end to uranium mining in 2010.

Most physicians have really only begun to educate themselves about ionizing radiation beyond limiting the use of x-rays. Just as physicians and the general public recognize that climate change in the form of global warming will have enormous impacts on health, a media campaign has been unleashed by the nuclear industry, claiming that nuclear power is "safe" and "green."[4] How does one sort out fact from fantasy?

In writing this book, we discovered that almost no one was neutral with respect to nuclear power; biases tended to extend to the research being conducted as well. This challenged us to look beyond the abstract or the conclusions of any one study and to review other work by the same author or to find out who commissioned the study, book, or news release. While this book was commissioned by PGS, the material presented herein is based upon good, evidence-based science. The scientific material has been cross-referenced by consultation with nuclear industry websites such as that of the World Nuclear Association. We wanted to produce a text where the facts could be trusted. Readers may disagree with one another and with us about what to do with the information. One reviewer commented, "The book is biased, but there is nothing wrong with the facts."

We give you permission to skim through the book, reading only the case studies at the beginning of most of the chapters. They tell the stories of companies, governments, and people that have been touched by ionizing radiation. We hope you also enjoy the book. If it inspires you to learn more or take action to prevent increasing personal or environmental ionizing radiation, our aim in writing the book will have been met.

Because we are doctors first and scientists second, our lens is different from that of physicists. We believe there are benefits to ionizing radiation but are more cautious where there are health implications. We make no apologies for our concern. We place no value in arguments that compare chemical or biological pollution from coal or gas with radioactive pollution and very little upon "dilution as the solution for pollution."[5] There is no room to compare the worst options to avoid what most scientists are calling an impending disaster.

The vast majority of us do not study ionizing radiation in school. At most, we receive an industry description of nuclear power and learn very little about the nuclear (atomic) bomb except that it was rumoured to have ended the Second World War. This book is an attempt to address that gap. It reviews sources of ionizing radiation, examines its health risks, and questions the environmental risks in continued development and use of this technology.

This is the doctor's voice.

Introduction

"ONIZING RADIATION, what is that?" a friend asks. It sounds too complicated, too scientific, to be of interest to the ordinary person. However, governments around the world are making decisions and creating policies on and about ionizing radiation that could affect all of us, not merely those of us alive today but also future generations. How can decisions be made if no one knows anything about it? Who informs our government leaders? One of the problems with technology is that the science can be so mind-boggling that the public leaves decisions about licensing and management up to the experts, usually within industry and with enormous conflicts of interest. Yet, now in the dawn of the twenty-first century, the world is awash with increasing ionizing radiation.

Radiation can itself be puzzling because there are so many kinds of radiation; light, radio waves, microwaves, and heat are examples of radiation, and, particularly, examples of *nonionizing radiation*. They are unable to break molecules apart because they don't have much energy. The electromagnetic waves of cell phones and Wi-Fi networks are also non-ionizing. Radiation is a form of energy transfer, "the emission of energy," according to the *Oxford English Dictionary*, "as electromagnetic waves or as moving particles."[1] While radiation may be non-ionizing or ionizing, this book is limited to high-energy, ionizing radiation.

Ionizing radiation has enough energy to break molecules apart and change the structure of substances. The broken parts of the molecules are called *ions*, electrically charged particles that carry either a positive or a negative charge. Positively charged ions will seek negatively charged ions and vice versa in order to bond and become neutral. The urge to bond can be weak, as in water, or extremely powerful, as in hydrochloric acid that so aggressively seeks neutrality that it breaks down food, digesting it in the stomach.

Ionizing radiation in the environment comes from the sun as cosmic rays, and from substances such as potassium, carbon, and uranium as alpha, beta, and gamma radiation. The sum of natural and artificial radiation in the environment is called *background radiation*. The damage that ionizing radiation causes to biological systems is additive, so the sum of a person's exposure is considered their *lifetime burden*. For most people, the greatest proportion of their lifetime burden of radioactive exposure will be limited to the use of x-rays, positron emission tomography (PET) scans, or computerized axial tomography (CT) scans, either for diagnostic or treatment purposes, background radiation being only a small part. On the other hand, there have been people who have suffered extreme effects of ionizing radiation. In all of its worst possible effects, ionizing radiation baffled victims of the atomic bombings of Hiroshima and Nagasaki in 1945 and spawned a permanent research facility, the Atomic Bomb Casualty Commission,[2] to focus on the effects of ionizing radiation at Hiroshima in Japan. Ionizing radiation from the radioactive plume from Chernobyl's 1986 nuclear disaster reached a swath of people in northwestern Europe; there was still a prohibition on UK sheep in 2009.[3] Monitoring the Pacific Ocean for ionizing radiation has been ramped up following the Fukushima meltdown in 2011.

Atomic or *nuclear* energy, the two terms will be used interchangeably, results from the breaking up of large atoms into a multitude of new smaller atoms releasing enormous amounts of energy as heat and ionizing radiation. The ionizing radiation released will be in the form of rays or particles. The principle of energy release was used in the nuclear bombs dropped in Japan, but the concept of taming atomic energy for a practically inexhaustible source of energy led to the "peaceful atom," a transformation of weapons-producing nuclear power plants into power plants to light homes and cities.[4]

Ionizing radiation is not without its problems, not the least of which is the regulation of its presence in the environment and the controversy over its safety. Most nuclear power regulatory bodies are composed almost entirely of physicists and engineers with experience in the nuclear power industry. While this creates bias in favour of the industry (after all, it paid their salaries for many years), there is another concern. Physicians and physicists tend to look at health risks differently. The nuclear industry finds an increase of one cancer per every one hundred workers in the nuclear industry "acceptable,"[5] while the physician says, "Just a minute—that is ten times the allowable risk in any other industry."[6] While the regulatory body may only see numbers on a page when it comes to the health risks of ionizing radiation, physicians and other health care providers sees the costs to the health care system, as well as the pain and suffering of the individual, their family, and their social network. The physicist thinks the increase is insignificant; the doctor and the epidemiologist agree that the increase is not only significant but, if preventable, entirely unacceptable.

This book is not a comprehensive review of every topic listed in its table of contents. In fact, we expect to disappoint readers who want more of a political treatise on the subject, critics who will insist the material is too biased, and scientists and technical specialists who will probably condemn the book as too superficial. There will also be those diametrically opposed individuals who will find the book too technical. Our intention has been to provide solid, basic information in a challenging field. There is useful educational material within the literature provided by the World Nuclear Association, the International Atomic Energy Agency, and the Canadian Nuclear Safety Commission, all promoters of ionizing radiation in industrial forms. Meanwhile, International Physicians for the Prevention of Nuclear War, Physicians for Global Survival, Bulletin of Atomic Scientists, and the Canadian Coalition for Nuclear Responsibility (CCNR) all provide sound scientific critique of ionizing radiation in the environment.

There are a few words in the book that could be stumbling blocks, especially when used before we get around to defining them: *radioactivity* and *radiation* are used throughout to mean "ionizing radiation," *isotopes* are different kinds of atoms of the same element, just as German shepherds and collies are different breeds of dogs, and *radioisotopes* are isotopes that are radioactive. There are several conventions with respect to atoms. We have chosen to use the name followed by the mass number (e.g., uranium-235) or the initial, also followed by the mass number, as in U-235.

Ionizing radiation is a difficult concept with much wider implications for human and environmental health. As global citizens, we have some responsibility to untangle the information, to make it more accessible—as much as such a difficult topic can be understood. The data, websites, and material included in this book should allow wider participation in the conversation about where ionizing radiation should fall within the spectrum of available technologies. We welcome feedback through www.pgs.ca.

CHAPTER 1

Historical Background

Mysterious Sickness[1]

AT THE SAME TIME that Columbus sailed to the "New World," in the 1490s, silver was discovered in the Cruel Mountains, so called for their harsh winter winds, in what is now the Czech Republic. In 1516, the area was named St. Joachimsthal by its founder, Count Stephan Schlick. While mining the silver, the miners found a strange mineral that stuck to their picks. Because it usually meant the end of silver in that vein, it was given the name *pechblende* ("bad luck rock"; "pitchblende") and tossed aside. Just across the border, in the German town of Schneeberg, Germans were finding the same black rock amongst the silver in their mine.

About fifteen years after the first mine shafts had been dug, miners on both sides of the border started suffering from what became known as Mountain Disease. The identifying features of the illness were a constant hacking cough and the spitting up of blood. Those affected usually died within a few months. Doctors were at a loss, trying to identify what was causing "the lungs to rot away," as one contemporary physician described it, and there was much speculation as to the cause. Nobody suspected the bad luck rock. In any case, Mountain Disease was soon forgotten when the silver ran out and famine and war devastated the land. In 1789, German pharmacist Martin Klaproth was experimenting with pitchblende and decided that it contained an as yet unnamed and unknown element. He called it "uranium," after a newly discovered planet, Uranus. Doctors had already speculated that Mountain Disease was caused by a gas and, when radon was isolated in 1898, it became suspect. It was not until the 1920s, however, that the mysterious, rotting-lung illness, now called Schneeberger lung disease, was identified as lung cancer.[2] By then it was generally accepted that it was caused by radon. Over the lifetime of the mine, up to 75 per cent of miners around St. Joachimsthal contracted lung cancer, depending upon which mine, and where in the mine, they worked.[3]

Discovery

In 1895, Wilhelm Röntgen (usually given the English spelling, *Roentgen*), using cathode rays, created *x-rays*, the first known form of ionizing radiation. The next year, 1896, Henri Becquerel accidentally discovered natural ionizing radiation. Becquerel was experimenting with pitchblende that glowed in the dark after exposure to sunlight (probably due to the ruthenium and some of the uranium salts in the ore). He reasoned that the same energy that caused the rock to glow in the dark would cast an image on photographic film. His experiments included exposure to light, so on a cloudy day, he was forced to postpone his experiment. Being an unusually tidy scientist, he cleaned up after himself and put the rock and a photographic plate with which he was working in a drawer. Days later, when he developed the plate, he was surprised to see the shape of the rock on the plate, exactly where it had been lying. Given that this had occurred in total darkness, it was clear to Becquerel that the process did not require sunlight, and that the minerals he was examining gave off rays by themselves.[4]

At the University of Paris, where she had registered in 1891, Marie Curie was floundering in the search for a topic for her Ph.D. thesis. Her supervisor suggested that she explore the phenomenon that Becquerel had discovered. She was the first to refer to the rays as being "radioactive." When she isolated uranium from her samples of pitchblende, she discovered that the remaining ore was highly radioactive, even more radioactive than the uranium itself. She began the painstaking process of isolating elements and was rewarded by finding both polonium and radium in 1898.[5]

Henri Becquerel, Marie Curie, and Pierre Curie, her fellow researcher and husband, were jointly awarded the Nobel Prize in Physics in 1903, Becquerel "in recognition of the extraordinary services he has rendered by his discovery of spontaneous radioactivity," and Pierre and Marie Curie "in recognition of the extraordinary services they have rendered by their joint researches on the radiation phenomena discovered by Professor Henri Becquerel."[6]

After the discovery of the new elements radium and polonium became known, Marie Curie was awarded the Nobel Prize in Chemistry in 1911. The award was notable in that she became the first person to receive it twice (only three other people have done so)[7] and because she was recognized for both the "isolation of radium" and the "study of the nature and compounds of this remarkable element."[8]

Early Effects on Health

Doctors eagerly adopted the new diagnostic and treatment possibilities that x-rays offered. As a result of their overenthusiasm, the earliest reported damage attributed directly to radiation was caused by x-rays and was evident within months of their discovery.[9] In 1896, an Austrian doctor reported he had severely burned his patient's back after treating a mole with x-rays.[10]

Thomas Edison, the US inventor of the light bulb, was also initially enthusiastic about x-rays and worked on the development of an x-ray focus tube. He lost his enthusiasm when one of his scientists, Clarence Dally, suffered radiation skin disease, which at first expressed itself as a dermatitis. This reddened and scaly skin damage progressed to skin cancer, which eventually spread (metastasized), resulting in his painful death in 1904. Edison himself suffered from sore eyes and skin rashes after experimenting with x-rays and later said, "Don't talk to me about x-rays, I am afraid of them."[11] Clearly, x-rays were not to be taken lightly, a fact reinforced when, years later, in 1926, H.J. Muller, a US geneticist, showed that x-ray damage to cells was hereditary in fruit flies.[12] This discovery would win him a Nobel Prize in 1946.

While x-rays were coming under scrutiny, natural sources of ionizing radiation fascinated the public. A vibrant industry grew up around the marketing of radium. It glowed in the dark and was known to have mysterious rays with equally mysterious powers. Unlike x-rays, it didn't cause immediate, visible damage.[13] In fact, radium was sold as beneficial to one's health in both Europe and the United States and radium-laced products were sold for health improvement, in cosmetics, and, ironically, for cancer prevention.[14]

During the mid-1920s, while in its heyday, slowly developing and sad events occurred that would bring this industry to a halt. The tragic story of the "radium girls" began to make headlines. Grace Fryer was one of the radium girls. She had begun working in a New Jersey radium dial factory in 1917. At the factory, the women used brushes dipped in a radium solution for painting glow-in-the-dark dials on clock faces. They were instructed to use their lips to form the brush to a point. Fryer wondered why her handkerchief glowed in the dark after she blew her nose, but she didn't worry about it. In fact, she and her co-workers—reassured that the material with which they were working was safe—had routinely painted their teeth and nails with radium to surprise their boyfriends at night.

Grace Fryer left the factory three years later, in 1920, to work at a bank, feeling young and healthy. Two years later, in 1922, she had excruciating pain in her mouth. Her teeth began falling out and x-rays showed that her jawbone was deteriorating (osteonecrosis). Initially thought to be "phossy jaw," previously identified in match girls and due to phosphorus, Dr. Theodor Blum, a dentist, called it "radium jaw." Years of denial and media manipulation ensued as U.S. Radium Corporation moved to protect its financial interests.

In 1927, Fryer and four former co-workers sued the corporation for $250,000 each. By this point, she had lost all her teeth, could not walk, and was unable to sit up without a back brace. In fact, it was clear to observers that the women were dying. U.S. Radium paid each woman a sum of $10,000 and $600 annually for as long as the woman lived. In addition, it agreed to cover all medical bills, past and future. It was a very cheap settlement even by 1920s standards; by the early 1930s, all five were dead. In the aftermath, it was discovered that the laboratory technicians and supervisors at the corporation's factory site had known about the dangers of radium and had used protective gloves and aprons to shield themselves.[15]

Radiation Protection Measures

As the mystery of radioactivity and its effects on human health unfolded, it left a trail of increasingly stringent regulations and protective guidelines and an entire alphabet soup of regulatory and scientific organizations. A short timeline of events can be drawn from the discovery of x-rays to the making of regulations: Roentgen discovered x-rays in 1895 and physicians immediately started using them. A year later, in 1896, x-ray dermatitis, a reddening of the skin with peeling, was described in medical literature in the United States and Europe. Radiation-caused dermatitis was puzzling because it was different from other rashes; it is very difficult or even impossible to heal because the *basal cells*, the cells in the skin that regenerate new skin, have been killed by the radiation. By 1900, US electrical engineer, Wolfram Fuchs, himself a martyr to the damaging effects of radiology, established the three principles that have become the tenets of radiation safety, principles of time, distance, and shielding for x-ray exposure and treatments. Patients were told to keep the exposure as short as possible, not to stand within twelve inches (30 cm) of the tube, and to coat their skin with Vaseline.[16]

Within six years of the discovery of x-rays, the link between x-rays and cancer had been established. At the time, radiologists would use their hands to focus the beam of the x-ray machine. They developed skin cancer at such an alarming rate that the connection between the two was obvious. The Röntgen Museum in Remscheid, Germany, retains the preserved and gruesome radiation-affected hand of radiologist Paul Drause as a reminder of the deleterious effects of ionizing radiation from x-rays on bone and skin.[17]

By using radiation experimentally for therapy for practically everything, it was likely that eventually radiation would be found useful for something. In fact, in 1896, basing his treatment on the already described skin changes, Emil Grubbe, a medical student in Chicago, used x-rays to treat a recurrent cancer of the breast.[18] By 1899, researchers in Sweden reported that cancer patients were being cured with the use of radioactivity.[19] The term *radiotherapy* was coined to refer to the use of radioactivity as a treatment.

At the turn of the twentieth century, both electricity and radiation were still novelties. Both technologies seemed to show promise for medical applications. Consequently, in 1902 an International Congress of Medical Electrology and Radiology was held in Switzerland to bring together people with direct and indirect experience with the new technologies. They discussed the injuries resulting from the use of x-rays, particularly acute and chronic dermatitis. They also discussed what could and could not be seen using x-rays; its use in diagnosing broken bones was obvious, but determining what could be seen in soft tissue, such as the lungs, was less straightforward.[20]

Acknowledging the usefulness of x-rays but interested in decreasing injuries, a dentist, William H. Rollins, after burning his hands while using x-rays for jaw and facial surgery, began experimenting with guinea pigs in 1904, eventually publishing a series of papers. He confirmed that the immediate ill effects of radiation from x-rays could be minimized by keeping the x-ray tube a certain distance from the patient's body, but he also advocated the wearing of lead-shielding for the parts of the body not involved in the x-ray and the use of protective clothing for the technician. With the exception of ending the practice of holding the x-ray tube against the patient's body, his research and resulting recommendations were ignored for decades.[21]

The problems of x-ray-related injuries plagued their use during the First World War, when an alarming increase in x-ray-related injuries resulted from the use of primitive, mobile, x-ray equipment in battlefield surgical units. Radiologists were forced to admit that the measurement of doses and exposure levels was inconsistent and poorly performed. To address this problem, the British Institute of Radiology invited radiologists from around the world to the first International Congress on Radiology (ICR) in London in 1925. It was at this first meeting of over five hundred medical radiation specialists—as much as one can be a "specialist" in newly discovered technology—that the International Commission on Radiation Units and Measurements (ICRU) was founded.[22]

A second ICR was held in Stockholm in 1928. Although the *roentgen*, a measurement of radiation, had been used to describe exposure for two decades, it was at this congress that it became the official measurement. The International X-Ray and Radium Protection Committee (IXRPC) was also founded to propose guidelines on radiation protection. By 1929, these guidelines forced the American Medical Society to officially condemn the use of x-rays for hair removal.[23] After that, in the early 1930s, following the death of a famous

New York sports personality and industrialist, Eben McBurney Byers, and the radium girls court settlement, the United States Food and Drug Administration (FDA) began cracking down on health and beauty products containing radium.

In 1934, the IXRPC established what it called a *tolerance dose*, a level of external radiation exposure considered "harmless." It was based upon levels of radiation that caused erythema (reddening of the skin).[24] Although the connection between radiation and cancer had been established in 1902, scientists assumed that the cancer effect was just a progression from the localized effect—if exposure stayed below levels causing erythema, it would not cause cancer. "Tolerance" described immediate, not long-term, effects.[25] Hence, because the long-term side effects of radiation exposure were unknown (and unsuspected), the tolerance dose was set at a level twenty-five times today's accepted occupational exposure level.[26]

A monument was erected in Hamburg, Germany, in 1936 to commemorate all those who had perished due to x-ray exposure. The medical profession wanted a symbolic way of parting with the unregulated, pioneering days of diagnostic x-rays and x-ray treatment by acknowledging the countless patients who unwittingly suffered the results of early experimentation. Several hundred medical workers who died from radiation damage are named on the monument as well.[27]

In 1946, in the United States, responding to scientists who had worked on the development of the atomic bomb and had experienced first hand some serious radiation effects (some had actually died), the National Committee on Radiation Protection (NCRP) was formed. The concept of tolerance dose was replaced with what it called the "maximum permissible dose." This new concept assumed that no level of radiation was completely safe but that the average person would be willing to take the risk of exposing him/herself to anything below the maximum permissible dose for employment or medical investigations. The NCRP also lowered the 1934 occupational standard by 75 per cent.[28]

After the Second World War, the ICR reconvened in London in 1950. In keeping with the example set by the NCRP in the United States, the IXRPC was renamed the International Commission on Radiological Protection (ICRP). It was restructured in order to include uses of radiation beyond the medical field. It retained a maximum permissible dose standard,[29] but reduced the annual exposure limit by 40 per cent.[30] In 1974, the ICRP created the *reference man* to use as a model against which to measure the effects of ionizing radiation.[31] A *reference woman* was based on the reference man by simply lowering the body mass, height, and weight. This has been heavily criticized because it fails to consider the different organs (breasts and uterus) and the different fat distribution in women's bodies. There has been no accommodation for a "reference child" or a "reference fetus"; in fact, allowable doses for children are calculated from those permissible for the reference man, and fetal exposure is only taken into account in radiation-controlled workplaces when a woman declares her pregnancy.[32]

Effect of Hiroshima and Nagasaki on Regulations

The dropping of atomic bombs on Hiroshima and Nagasaki in Japan in 1945 profoundly altered the ways in which the world viewed ionizing radiation. After a media blackout imposed by the United States, the Atomic Bomb Casualty Commission

(ABCC) began five years after the bombings to document the effects of radiation on the survivors. It began the Life Span Study, a longitudinal, epidemiological study that followed and still follows the lives of atomic bomb survivors, using the effects of estimated radiation doses to generate data on cancer incidence, cancer mortality, and noncancer effects. The ABCC was transformed into the Radiation Effects Research Foundation (RERF) in 1975 with increased Japanese leadership. One interesting finding was that, in the survivors registered, the cancer incidence paralleled the expected incidence in the normal population; that is, the survivors did not get cancer any earlier than nonexposed people. They got cancer at the same ages as nonexposed people, but the incidence was greater, statistically relevant but less than expected, only 1 or 2 per cent higher for each type of cancer.[33]

There are problems with the ABCC/RERF data. The studies were started late—the earliest records are from 1947, but the commission didn't really get going until 1950, after the immediate and subsequent deaths from illness (and infections from destroyed immunity) had occurred. They were faced with enormous suspicion by Japanese victims, who thought they were being treated like "guinea pigs,"[34] so it is unclear who came forward to be examined. The ABCC also refused to use any material that the Japanese physicians submitted and, according to one physician, confiscated the material. Speaking at the International Physicians for the Prevention of Nuclear War (IPPNW) Congress in Japan in 2012, one elderly physician reported that he had been collecting blood samples, had found anomalies in blood proteins, and, to that date, was sad that he had given his material to the ABCC.[35] In fact, the ABCC stated that "the death rate from all causes except cancer had returned to 'normal' and the cancer deaths were too few to cause alarm."[36]

Furthermore, the bombs dropped on Hiroshima and Nagasaki delivered a single, external, high-energy blast of radiation from untold numbers of short-lived, artificially created, new radioisotopes. People were affected differently depending upon their distance from the blast and their exposure (for example, whether they were shielded behind a wall or not). The radiation and the radioactive isotopes were unique. The radiation from the bombs was substantially different than the usual low-dose and/or chronic exposure from nuclear power plants or technical devices. Even though the resulting health data is not comparable, remarkably, both the ICRP and the United Nations Scientific Committee on the Effects of Atomic Radiation (UNSCEAR) continue to base standards for allowable exposure from nuclear plant emissions upon the ABCC/RERF studies of Hiroshima victims.

Radiation Safety

When the ICRP convened in Copenhagen in 1953, it tackled radiation safety. Standards needed to be established not only for radiation workers but for the general public as well. The commission somewhat arbitrarily recommended that the general population should not be exposed to more than one-tenth of the allowable exposure for a radiation worker. It reiterated that no level of radiation was absolutely safe and acknowledged the difficulty of quantifying a limit that represented a negligible risk. A new unit of measurement known as the *rad*, a calculation of the absorbed dose of radiation, was established to recognize that the same amounts of radiation posed a different level of risk to different human organs. An effort was also made at this meeting to distinguish between the maximal permissible dose in water and air.[37]

Four years later, in 1957, the ICRP recommended that workplaces dealing with radiation be divided into controlled and uncontrolled areas. A *controlled area* was defined as an area having a supervisory radiation safety officer. An *uncontrolled area* was defined as an area without supervision. Controlled areas have greater risk of exposure to radiation. The commission recommended that the safety limit for a person working in an uncontrolled area should be the same as for the general public (i.e., one-tenth of the allowable limit for workers). Because animal tests had previously demonstrated how radiosensitive embryos are, it also recommended that pregnant women avoid occupational radiation exposure entirely.[38]

Although the ICRP initially focused on the immediate effects of radiation, by 1966 it had become more concerned about delayed outcomes such as cancer and genetic effects. It introduced the Linear No-Threshold Model, arguing that there is no lower limit for the effects of radiation. In other words, this model proposed that the risk for, and probability of developing, cancer increases with every dose above zero. In 1977, the commission introduced three principles of protection:

1) No practice shall be adopted unless its introduction produces a positive net benefit.
2) All exposures shall be kept "as low as reasonably achievable" (which became known as the ALARA principle), economic and social factors being taken into account.

3) The doses to individuals shall not exceed the limits recommended for the appropriate circumstances by the commission.[39]

That same year the ICRP was also the first to mention the protection of nonhuman species by extrapolating, without any evidence, that if humans were kept safe, then other species were likely safe as well.[40] Superficially, this might seem reasonable, but given that different tissues respond differently to radiation, it is quite likely that different organisms also respond differently.

During the 1980s, the ICRP was criticized for underestimating the risks of exposure for some people under specific circumstances (for example, industrial use of x-rays, transportation of enriched uranium, and many others) because its recommendations were too general. In response to this, by 2006, it had developed thirty different recommendations to account for different situations and circumstances. Any dose of radiation was recognized to increase the risk of cancer, teratogenic effects on the fetus,[41] and genetic disease. The recommended doses are simply the lowest doses that meet the criteria of being "as low as reasonably achievable, taking economic and social factors in account," not the doses that provide the healthiest human and environmental results.[42]

Meanwhile, in the United States, beginning in the 1960s, the National Academies of Science, through its Committee on the Biological Effects of Ionizing Radiation (BEIR), started a series of literature reviews, published irregularly as new information became available. Usually these reviews stopped short of assigning danger to naturally occurring background radiation, but in 2005, the seventh report on the

Biological Effects of Ionizing Radiation (BEIR VII) was notable. It concluded that there is no safe level of radiation and that even background radiation is a cause of cancer.[43]

In 1956, the World Health Organization (WHO) issued a warning: "Genetic heritage is the most precious property of human beings. . . . As experts we affirm that the health of future generations is threatened by increasing development of the atomic industry and sources of radiation." After a second cautionary statement, the International Atomic Energy Agency (IAEA) effectively gagged WHO on matters of health in the presence of ionizing radiation. WHO subsequently agreed to "consult the IAEA" before issuing statements on anything in which the IAEA might have an interest. Subsequently, WHO has released no critical statements about ionizing radiation since the agreement was signed in 1959. In 2003, British radiation biologist Keith Baverstock was, according to the *Guardian*, "sacked [from WHO] after expressing concern that new epidemiological evidence . . . indicated that the current risk models for nuclear radiation were understating the real hazards."[44]

Nuclear Safety in Canada

In Canada, the Canadian Nuclear Safety Commission (CNSC) is responsible for setting safety limits for radiation exposure for workers in hospitals and nuclear power plants, as well as for the general public. The limit for workers, if reached, creates an increased occupational risk of cancer that is logarithmically greater than the allowable occupational risk of disease in any other industry. Aside from concerns about these standards, the independence of the CNSC from industry and government interference became questionable when the Canadian government ordered that a reactor be restarted over the safety objections of the then commissioner, Linda Keen. The Chalk River nuclear reactor, located about 180 km upriver from Ottawa, Ontario, had twice failed to provide backup pumps for one of its water systems. Having already issued a warning, as head of the CNSC, Keen ordered that the reactor be shut down until such time as it passed its inspection. Not only did the federal government interfere with the decision of the safety commission but, on January 15, 2008, it extended its power to fire Ms. Keen.[45]

Most people who work where they may be exposed to radiation are required to wear *dosimeters*, small badges that track personal exposure to radiation. The badges are turned in regularly so that the amount of radiation to which an individual worker has been exposed can be known over time. The badges are only useful in so far as they are worn, or worn unhindered. Following the Fukushima nuclear power accident of March 11, 2012, during the clean-up, some workers were reportedly supplied with lead coverings for their dosimeters by their contractor so that the recordings would be low and the employees could stay at work.[46] Aside from measuring exposure so that workers can take protective measures, other measures are also taken—or should be taken—in hospitals and in dentists' offices, such as having the technician step behind a leaded wall or the wearing of lead aprons and thyroid protection. The walls and windows surrounding an x-ray department are also heavily shielded.

The Canadian government has maintained and updated a database on workers' exposure to radiation since the 1950s.[47]

It has used this to detect trends and respond with new safety standards when necessary.

Closing Comments

Ionizing radiation was a complete mystery when it was discovered at the end of the 1800s. It was quickly embraced as a tool with hundreds of applications, many of them proving false or too dangerous to pursue further. Through its exploration, the insides of atoms have been revealed. The history of ionizing radiation is also the history of radiation protection; trying to understand the health effects of ionizing radiation and setting limits on allowable exposure for patients, employees, or anyone who might be exposed. Both industry and regulators use the ALARA principle to guide standards. This means that regulators negotiate with industry in setting these limits, not with health care professionals. From its discovery and first use, regulators have lowered the maximal permissible exposures to ionizing radiation at least seven times. Yet, many health care workers still wonder, did they get it right this time?

CHAPTER 2

Radiation Science

FOR READERS who don't have a science background, this chapter probably contains more scientific and medical jargon than used in the average lifetime. Read it or skim the titles, and come back to it if you need to refresh your memory.

Atoms

Atomic energy comes from atoms. An *atom* is the smallest piece of an element[1] that still has the properties of that element. An atom of hydrogen is infinitesimally small, but it will still be hydrogen. An atom is like a miniature solar system with a centre, its nucleus, and electrons that are like planets circling the nucleus.

Atoms (and elements) are designated by symbols—one or two letters, usually from the beginning of their names (if two letters, the first is capitalized). For example, He is the symbol for helium, U for uranium, and Pu for plutonium. The *mass* (weight)[2] of an atom is the sum of particles within its nucleus and is considered to be its atomic weight. (Electrons are so small that they contribute practically no weight.) Helium has two neutrons and two protons (Figure 2.1) in its nucleus, so its mass or atomic weight is four. The uniqueness of an element is based upon the number of protons in its nucleus—its *atomic number*. Since the number of protons is balanced by the number of electrons, the atomic number also represents the number of electrons circling the nucleus. Since helium has two protons in its nucleus, its atomic number is two and two electrons orbit the nucleus. Protons are positively charged; electrons are negatively charged. A *neutron* is made up of a proton plus an electron and is neutrally charged. Neutrons behave like a type of glue to keep positively charged protons together as nuclei become larger.

Large atoms are almost always represented by their initials plus their mass number because they usually have a number of

Figure 2.1: Helium Atom

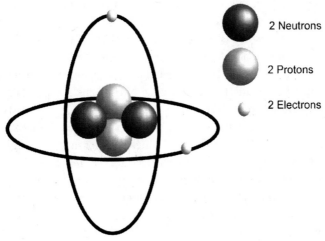

2 Neutrons

2 Protons

2 Electrons

Source: F. Oelck

isotopes. An *isotope* is another variation of the same element, with the same unique atomic number but a different mass. Uranium has three naturally occurring isotopes, each with an atomic number of 92, but since their mass is different, they are identified by their mass: uranium-238 ("depleted" uranium), uranium-235 (used in nuclear reactors and bombs), and uranium-234 (a natural decay product of uranium-238). The isotopes share the same physical properties of uranium—the same melting point, the same types of compounds, etc., but each has different atomic properties. Their atomic weight is not unique; besides uranium-238, plutonium-238 and neptunium-238 also exist. What distinguishes them as separate elements is their unique atomic numbers: 92 for uranium, 93 for neptunium, and 94 for plutonium. When an isotope is radioactive, it may be referred to as a *radioisotope*. Since all isotopes of uranium and plutonium are radioactive, the term is redundant, but in the case of an element like carbon, carbon-14 is the only radioisotope of carbon.

An element that changes into another element is *radioactive* and is considered to be *unstable*; elements that do not change are *stable*. Helium is stable. Some isotopes of an element can be radioactive even though the element is generally considered stable like, for example, carbon. Carbon exists in its most plentiful form as carbon-12, but one-trillionth of natural carbon is carbon-14, which is radioactive. Although radioactivity is associated with larger nuclei, even little hydrogen has a naturally occurring radioisotope, tritium.[3] On the other hand, the largest element with a stable isotope is bismuth-208 with an atomic number of 83; every element whose atomic number is greater than 83 is radioactive.

Radioactive Decay

When a radioactive atom changes into another element, it is said to have undergone *radioactive decay*, giving off energy during the change. The mass and/or energy released during radioactive decay is *atomic energy* or *nuclear energy* (for most purposes, the two terms can be used interchangeably), identified and measured as *radioactivity*. The atom undergoing decay is called the *parent*, and the new atom that is created is referred to, also interchangeably, as the *decay product*, the *daughter*, or the *progeny*.

When the parent potassium-40 decays to progeny calcium-40, or carbon-14 becomes nitrogen-14, their nuclei have reached stable states. When the nuclei of large atoms decay, they usually have many steps forming many unstable isotopes before they reach a stable state. The series of decay steps is referred to as a *radioactive decay chain* or simply *decay chain*. Uranium-235 has a long, complicated decay chain. The first step produces thorium-231 (Th-231) and an alpha particle. In the second step, thorium-231 releases a beta particle and becomes protactinium-231 (Pa-231), which is also unstable (see Figure 2.2, illustrating the first two decay steps). The weights of thorium and protactinium are the same, but the atomic numbers, the number of protons in the nucleus that identify the element as different, are 90 for thorium and 91 for protactinium. As U-235 goes through its entire decay chain, thirteen separate elements will be created before ending at lead-207 (Pb-207). Another example, radon-222, passes through seven discrete steps before it becomes lead-206. Each step within radioactive decay is absolutely inevitable and each radioactive element will have its own distinctive

Figure 2.2: First Two Steps of Uranium–235 Decay

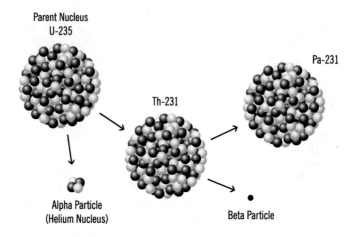

Source: Michael Webb, Alchemy Design

decay or decay chain. Unlike water running downhill, no dam can be built to stop it.

Types of Radiation

Of the radiation particles, *alpha particles* are the largest, consisting of two protons and two neutrons, equivalent to the nucleus of a helium atom. They do not travel very far and can be stopped easily by a sheet of paper. Before scientists recognized the risk of internal radiation in the form of aerosolized particles, it was thought that alpha particles were harmless, since, at most, they can penetrate only two levels of cells on the surface of the human body. In fact, as we will see, alpha particles are considered to be twenty times as dangerous to living cells as gamma rays.

Beta particles, on the other hand, are high-speed electrons with practically no mass. They are ejected by the nucleus when a neutron becomes a proton. They are faster than alpha particles and can penetrate paper and layers of skin, for example, passing entirely through a hand, but are stopped by plastic or wood. Beta particles are usually negatively charged, but a few elements decay by emitting *positrons*, positively charged particles with similar characteristics.

Gamma rays are packets of energy released, usually but not always, when some alpha or beta particles are released. They are *photons*, pulses or packets of energy, not particles. They are similar to x-rays and can be stopped only by a thick layer of a dense substance such as lead or concrete (see Figure 2.3). A few elements that have use in medicine are *pure gamma emitters* such as technetium-99m (see Chapter 4).

Figure 2.3: Penetrating Ability of Various Types of Radiation

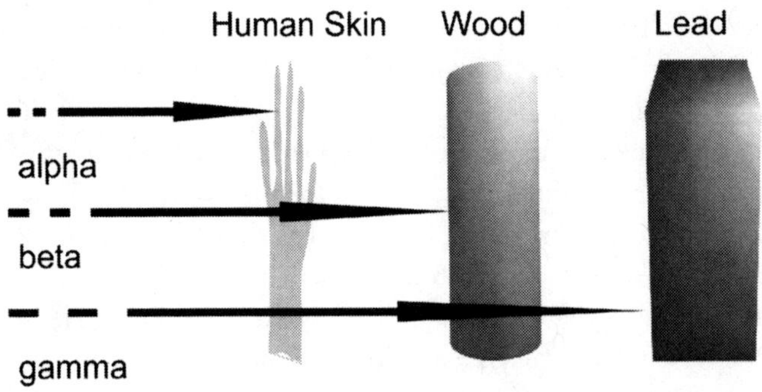

Human Skin   Wood   Lead

alpha

beta

gamma

Source: F. Oelck

The *neutron* is a third radioactive decay particle. It doesn't occur naturally. A neutron has to be forced out of its nucleus. It is considered *ionizing* because its neutrality (with respect to electrical charge) allows it to penetrate deeply into materials (including biological systems) where it changes into a proton and an electron, both of which are ionizing. Neutron radiation is highly lethal. In the 1950s, contemplation of an extremely lethal neutron bomb, which destroyed living things but had minimal effect upon inanimate buildings and objects, led scientist and science fiction author Isaac Asimov to call it "desirable to those who worry about property and hold life cheap."[4]

Half-Life

Half-life (indicated by $t^{1/2}$) refers to the amount of time required for half of the atoms in a radioactive material to decay, to become something else. Each radioactive element has its characteristic half-life. For example, iodine-131 (an artificially created radioactive isotope of normal iodine-127) has a half-life of eight days and is used to treat some types of thyroid cancer. (It is also present in fallout from nuclear power plant accidents.) The cancer patient is given a drink or a capsule containing a dose calculated on the sensitivity of the cancer to radioactivity. During the first week after drinking it, patients are so radioactive that they, their clothing, and body waste are isolated. They are monitored until the radioactive level within them is safe enough for them to leave the hospital, but they are told to avoid pregnant women and warned that they may set off security alarms in airports up to three months after the treatment, equivalent to about ten half-lives. Decay can be visualized as simply a quantity of atoms.[5] By the end of one half-life, half of radioactive iodine-131 will have changed into another element, stable xenon-131, by emitting both gamma rays and beta particles. Hence, by eight days, half of the iodine-131 has changed into xenon-131. By sixteen days, two half-lives, only one-quarter of it remains. By ten half-lives, the length of time required for a radioactive element to completely "disappear" into its decay product, all of the iodine-131 will have effectively become xenon-131. The process never reverses.

Half-lives are specific for each radioactive element but vary enormously between different elements. Uranium-238 has a half-life of almost 4.5 billion years, but one of its decay products, astatine-219, blinks into and out of existence in less than a second. Radium-226 takes 1,600 years for one half-life to occur, but half of its radon-222 daughters will have changed into polonium-218 after three and a half days.

Tailings, left-over waste from uranium mines, contain all of the progeny of uranium that are produced during all of the steps of its decay. Each element in this decay chain is radioactive. These include thorium, radium, polonium, radioactive lead, and lead itself. In the undisturbed rock in nature, all of the elements in the decay chain exist in equilibrium, very slowly cascading into the stable end product, lead-206 (Figure 2.4). This equilibrium is disrupted by the mining and milling process.

Background Radiation

Background radiation has both natural and artificial sources. The natural sources are food, gamma rays from the sun, and

Figure 2.4: Uranium–238 Decay Chain

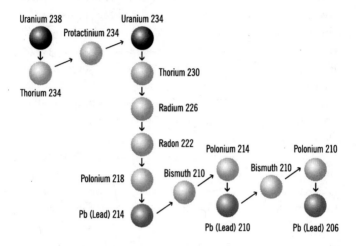

Source: Michael Webb, Alchemy Design

the alpha particles from radon gas. Food contains natural radioactive elements with different types of emissions. Potassium, an essential element for animals including humans, contains 0.0117 per cent potassium-40. Potassium-40 decays to calcium-40 by beta emission (or to argon-40 by electron capture).[6] With a long half-life of 1.25 billion years, it enters the body and is almost as quickly removed through the kidneys along with normal potassium. Bananas and other foods claiming to be "high in potassium" will have very tiny amounts of increased radioactivity.

Radioactive carbon-14 also derives from natural sources. It is found in trace amounts in all living things, including the food we eat and in the carbon dioxide we breathe. It decays into nitrogen-14, with a half-life of about 5,700 years. Its presence is used for *carbon dating*—the detective work of determining the age of ancient artifacts or disproving fakes.

While background radiation is all around us, for most people, medical imaging procedures such as x-rays, dental x-rays, CT scans, and PET scans comprise a far greater proportion of their lifetime radiation exposure. People wishing to decrease their personal burden of radioactive damage should balance the risk of exposure by the clinical information gained by the medical procedure. For example, if an x-ray will not change the treatment, a person might consider avoiding the radiation exposure.

Units of Measurement

A Greek philosopher named Democritus described atoms in 460 BCE as the smallest indivisible particle of a substance.[7] Centuries later, in 1803, John Dalton published his atomic theory.[8] The nineteenth century became a heyday for chemists as scientists combined atoms into molecules and separated them again. They made, and analyzed, thousands of different compounds and elements. A century later, the scientific world was equally charged over the discovery of ionizing radiation. The atom itself could be explored! Now standardized ways to describe the different findings were needed.

The entire system of units evolved piecemeal as characteristics of radiation became known. At first it was thought that a description of the type of radiation—alpha, beta, gamma, neutrons, positrons, and x-rays—was enough to describe the effect of the radiation. Very soon it was apparent that each type of radiation could have different levels of energy. Later, as physicians and biologists worked with radiation, they discovered that different types of tissue (e.g., liver, bone, or brain) showed different effects at the same dose. Furthermore, different types of cells in the same organ behaved differently upon receiving the same dose. Their absorbed dose should have been the same but clearly the damage was different. Something like a damage factor was needed.

In summary, the measurement of radioactivity is complex. There are different types of radiation—alpha, beta, gamma—and each of these is different from the others in terms of physical characteristics. Similar particles from different elements will have different energy levels distinctive for each element. For example, two elements, iodine-131 and rhenium-188, used to treat cancer, emit both beta and gamma radiation. Iodine-131 emits very strong gamma radiation. It is extremely efficient at killing off cancer cells in the thyroid. The weak beta radiation of I-131 is largely ignored. Meanwhile, rhenium-188 emits high-energy beta particles that are used to treat liver cancer and its weak gamma radiation is ignored.

It is probably intuitive that the amount of biological damage produced by a stronger particle or ray will be greater than that inflicted by a weaker particle or ray. Shortening exposure times is also associated with less radioactively induced damage, a fact put to use, for example, when people limit their exposure to sunlight to decrease the risk of skin cancer.

There are over three thousand radioactive isotopes existing naturally or artificially. Since each one will exhibit a characteristic radioactive decay, this represents an enormous range of energy from atoms. Quantifying something with so many variables has resulted in the following complicated series of measurements:[9]

The amount of radioactivity emitted by a source is measured in becquerels and curies. One *becquerel* (Bq) is one atomic disintegration (or decay) per second. Since this is an extremely small unit, prefixes are commonly used. For example, potassium in the human body emits four thousand such disintegrations per second, so its activity is described as 4 kBq (kilobequerels). A *curie* (Ci) is an older designation named in honour of Marie Curie and based on the radioactivity of one gram of radium-266: 1 Ci = 3.7 x 10^{10} Bq = 37 GBq (gigabecquerels).

The amount of energy that is absorbed by a target is the *absorbed dose*, measured in grays and roentgens (older values): one *gray* (Gy) equals one joule per kilogram of tissue.[10] One gray equals one hundred rad. The *rad* is an older, less frequently used measuring unit defined as the dose causing one hundred ergs[11] of energy to be absorbed by one gram of matter. Grays can also be referred to as milligrays. For example, the absorbed dose of a CT scan of the head would be in the range of 44 mGy. One *roentgen* (R) is 2.58 x 10^4 C/kg (coulombs per kilogram) in dry air.[12] It is generally used only in the context of x-rays or gamma rays and measures the ionization in the air caused by a given dose.

In order to measure the biological damage that occurs as a result of exposure to a particular type of radiation, the absorbed dose is modified by a *radiation weighting factor* (RWF) to create the equivalent dose in sieverts. The RWF for beta particles is one, that of alpha particles is twenty, and that of gamma rays (x-rays) is one. One *sievert* (Sv) is 1 Gy x RWF. For example, the CT scan of the head that gives an absorbed dose of 44 mGy likely has an equivalent dose of 44 mSv. (Since the RWF for x-rays is one.) One sievert equals one hundred rem, a *rem* being an older unit that is occasionally used.

The sievert is actually a fairly large unit. One of the physicists working on the Manhattan Project accidentally exposed himself to an estimated dose of 5.1 Sv[13] and died shortly thereafter. The most commonly seen measurement is the *millisievert* (mSv) (1 mSv = 0.001 Sv) or the much smaller *microsievert* (µSv) (1µSv = 0.000001 Sv). For example, average worldwide background radiation equals 2.4 mSv per year. Meanwhile, a chest x-ray delivers about 0.1 mSv, and a single airport security screening delivers, at the most, 0.03 µSv (or 0.0003 mSv).

Sievert measurement is not an exact science. Exposure in sieverts is derived from knowledge of the exact types of radiation, the amounts of energy of each, plus the distance between the source and the target, being the person or object. Attempts to measure annual radiation exposures, especially for workers, may involve so many variables that the margin of error can be greater than 50 per cent!

When radioactivity is used in medicine, there are two additional ways in which the dose is measured, effective dose and committed dose:

Different parts of a human body respond differently to radioactivity. The brain and nervous system are relatively resistant, while bone marrow is quite sensitive. The amount of radioactivity from the source, the emitted dose, and the absorbed dose are modified by the varying sensitivities of different tissues in the body to produce the *effective dose*. The International Commission on Radiological Protection (ICRP) is responsible for determining specific tissue doses. Going back to the example of the CT scan of the head, since the scan is targeting nerve cells in the brain, which are relatively resistant to the damaging effects of radioactivity, the effective dose is much lower than the absorbed dose of 44 mSv. Instead, the effective dose is merely 1–2 mSv.

A *committed dose* of radioactivity represents the total, long-term biological dose received from radioactive materials inserted or embedded in the human body. For example, a committed dose can be calculated for *brachytherapy*, a medical procedure where beads of radioactive material, such as cobalt-60, cesium-137, or iridium-192, are inserted into a cancerous tumour. Committed dose is a function of the types of radiation given off, the half-life of the radioactive material, and the amount of time it is expected to remain in the body and is usually calculated for a fifty-year period following insertion of the radioactive material. Committed dose also applies to shrapnel embedded in soldiers or civilians following injury by uranium-238 (depleted uranium) weaponry. In fact, any radioactive substance that remains in a human body will deliver a committed dose of radioactivity.

These measurements are used to determine occupational and civilian safe exposure levels, keeping in mind that no level is considered without some risk. Individuals can also calculate their lifetime exposures knowing background and voluntary medical investigations or industry exposures. The aforementioned potassium will, over fifty years, expose a person to 0.1 microsieverts, which would be part of the background level to which they were exposed. Average background radioactivity in the world is around 2.4 millisieverts per year (mSv/yr), although in some places it can be as high as 10–20 mSv/yr, depending upon location and geography.[14] For example, Colorado in the United States is at a relatively high elevation and thus exposed to more radiation from the sun than lower-lying places. Valleys may "concentrate" the heavy gas, radon, arising from the granite in the mountains around them, while wind-swept plains would have lower levels of naturally occurring radon. Canadian exposure varies but is usually slightly above 3.0 mSv/yr because of the amount of granite in the Canadian Shield. Higher levels of naturally occurring radiation appear to be linked to higher levels of chromosomal errors.[15]

The current exposure limit for Canadian and US workers is set at a maximum of 50 mSv for any single year and a total of 100 mSv over five years.[16] The Canadian Nuclear Safety Commission says that most workers receive no more than the public, where the standard is 1 mSv per year in excess of background radiation. In spite of a flurry of objections, the Japanese government has accepted a safety standard of 20 mSv per year for school children in the Fukushima nuclear power plant disaster area, a case of changing the standards to match the circumstances.[17]

Closing Comments

Atoms are unique from one another, each a representative of a specific element. Large atoms, in particular, may be unstable and discharge mass and/or energy in an attempt to reach a stable state. This release of energy or mass is interpreted as radioactivity.

As more and more has been learned about ionizing radiation, it has been a challenge to measure it, contain it, and regulate it. The most frequently used measurements are becquerels for the amount of radiation released by a radioactive source and millisieverts for the amount of radiation received by a target, but these are not exact, the results being complicated by the exact natures of both the source and the target.

CHAPTER 3

Radiation and the Human Body

ONIZING RADIATION causes oxidative stress to body cells. Many people are familiar with the marketing of anti-oxidant supplements like vitamin C or vitamin E—also contained in some foods such as broccoli, garlic, and raspberries—to prevent cancer. More than any other source, ionizing radiation creates ions that cause oxidative damage.

Oxidative damage can be defined by using the example of what happens when a gamma ray hits a cell. When the ray hits the cell, it breaks up molecules. Oxygen and other negatively charged ions are released. Cells move waste and food across their walls, which are thin skins that have many small positively and negatively charged holes, by an exchange of ions. The negative oxygen ions interrupt normal cellular function because they are attracted to and plug up positively charged spots along the cell's membrane. In 1972, a Canadian, Dr. Abram Petkau, was able to show, using artificial cell membranes, that a low dose of radiation is as efficient at rupturing cell walls as a high dose. When a cell's membrane is exposed to a high dose of radiation (15.5 sieverts [Sv] per hour), the large number of oxidative ions can utterly destroy the membrane, but they are just as likely to recombine with one another, leaving the membrane intact. When the same type of membrane is exposed to a low dose over a longer period of time, the ions so formed are spread out and are unlikely to react with one another, each combining instead with either a cellular structure, or, in Petkau's experiment, the cell's membrane.[1]

Internal versus External Radiation

There are two kinds of ionizing radiation with respect to the human body: internal and external radiation. When an individual has an x-ray, he or she is exposed to external radiation. *External radiation* leaves a path of ions and altered molecules along its route through the body but no radiation remains in the person's body. *Internal radiation* occurs whenever a natural or synthetic radioactive element gains access to the human body by ingestion, skin absorption, or, in the case of medical use, injection or implantation. The radiation continues as long as the element remains radioactive. For example, inhaled uranium particles or radon gas result in internal radiation. In the case of the radium girls (Chapter 1), as long as the radium remained in the paint pots or on the dials of the clocks, it was an *external emitter* and fairly harmless. However, when Grace Fryer and the other workers licked their paintbrushes, the external radiation became internal radiation and the radium became an *internal emitter*.

The Radiation Weighting Factor

The radiation weighting factor (RWF), mentioned in the previous chapter, is used to calculate biological risk. Alpha particle radiation is comparatively harmless when coming from an external source because the particles are too bulky and don't have enough energy to force themselves through human skin. In the past, promoters of uranium mining have capitalized on this feature of alpha particle radiation by demonstrating the "safety" of radioactive ore to unsophisticated villagers in northern Saskatchewan, Mali, and Tanzania by applying yellowcake to the skin of their hands and cheeks.[2]

Alpha particles are much more dangerous when they are internal emitters. Even though they cannot travel very far, their size[3] allows them to do large amounts of damage, breaking up enzymes and structural molecules, all the time creating ions along their tracks. The amount of damage makes it very difficult or even impossible for the cells to repair themselves. Beta particle rays are much smaller but can travel farther, dispersing the same type of damage along a narrow trail, which, because of its smaller size, is easier to repair.[4] Beta particles cause more damage if they are internalized because they cannot penetrate the body from the outside very deeply. Gamma radiation leaves a longer trail of damage than beta particles but doesn't remain in the body. Consequently, the RWF of beta and gamma radiation is one, whereas that of alpha radiation is twenty. Thus, alpha radiation is considered to be twenty times more damaging within biological tissue than either beta or gamma radiation.

As previously mentioned, biological risk is measured using *sieverts* (as well as the older *rem* unit, which stands for Roentgen equivalent man) and is obtained by multiplying the RWF by the exposure in rads or grays (Gy). For example, if a person received 0.2 Gy of internal alpha radiation (RWF of 20), the biological risk would be: 20 x 0.2 = 4 sieverts (4000 millisieverts [mSv]). A dose of 5000 mSv is considered lethal within one month to 50 per cent of those exposed. Four thousand millisieverts is a large dose. If a person was exposed to 0.2 Gy of internal beta radiation (RWF of 1), their biological risk would be much lower: $1 \times 0.2 = 0.2$ sievert (200 mSv). A dose of 200 mSv could potentially cause nausea and fatigue. An exposure of one sievert over a year is estimated to increase the risk of cancer by 5.5 per cent.

Ionizing radiation has been used deliberately as a murder weapon, one of the most recent victims being Alexander Litvinenko, a former Russian secret service officer. On November 1, 2006, an estimated ten micrograms of polonium-210 were slipped into his tea. He developed vomiting and diarrhea almost immediately. His hair fell out shortly thereafter and three days later, when he was admitted to hospital, his bone marrow was failing to produce either white or red blood cells. The cause was difficult to detect. Not only was the quantity extremely small but polonium-210[5] is an alpha emitter and most radiation detectors (Geiger counters) cannot detect alpha radiation. The amount of polonium-210 given to Litvinenko had a radioactivity dose of two gigabecquerels, about two hundred times a lethal dose. His committed dose was calculated to be about 20 Sv. Polonium-210 is an extremely potent poison; the ten micrograms he ingested was smaller than the ink in the period at the end of this sentence.[6]

How Radiation Damages Cells

The oxidative stress caused by radiation can damage any molecular component within a cell. Damage to the *cytoplasm*, the content of the cell excluding the nucleus,[7] and cell membrane can be repaired by the cell itself, provided the dose or doses are low enough that the cell's defences are not overwhelmed. More significant is the damage to *deoxyribonucleic acid* (DNA), the genetic material that directs cells' activities, and *mitochondria*, the industrial centres of human cells.

When a radioactive particle or energy ray strikes any molecule, it can break it up—like splitting water into oxygen and hydrogen. When it collides with a DNA molecule, only one strand of its double helix might be broken. A single-strand break is usually repaired easily because the cell simply uses the opposite strand as a template. However, if the radioactive particle causes a complete break of both strands of DNA, the cell can't perform the repair. The template is lost. The fragments may lose one another or even connect by linking to other damaged sites (see Figure 3.1). The resulting damage is a witness to the destructive activity of the radiation.[8]

Damage to cells has a wide range of implications. The cell might repair itself (which occurs frequently during a single-strand break in DNA), or it might die. Neither of these alternatives is likely to affect the health of a person if only one or two cells are involved. On the other hand, the cell might continue to survive with a disrepair or mutation, which can have different implications. The mutated cell might either simply remain dormant, and have no negative impact, or it might multiply, which could eventually lead to cancer. Finally, the mutation might be passed on to offspring (if it involves egg, sperm, or reproductive capacity) and trigger genetic defects either in the next generation or carried as a recessive gene to express itself in future generations, an effect that has been seen in shorter-lived birds and animals around Chernobyl.[9]

Figure 3.1: Broken DNA

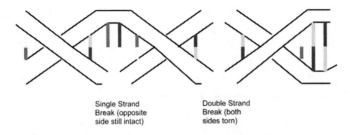

Single versus double strand break

Single Strand
Break (opposite
side still intact)

Double Strand
Break (both
sides torn)

Source: F. Oelck

The Impact of Radiation on Our Health[10]

There is no "zero" for radiation exposure; we are all exposed to natural background radiation to some degree. The sun shines on us, rocks emit ionizing radiation, and potassium-40 and carbon-14 are in our food, so we are exposed to both external and internal radiation. Every single dose is additive throughout our lifetimes. On the other hand, artificially created ionizing radiation has been entering the environment through nuclear weapons testing, nuclear power plants, and a host of health technologies. Long-term, low-dose exposure usually doesn't cause any obvious symptoms, but even though no immediate effect is evident, eventually some people will develop systemic illnesses, *autoimmune disorders* (where the body attacks itself, such as in systemic lupus), cancer, and genetic effects. Other effects may be even more insidious;

some studies have indicated trends, not statistically significant, that deserve more follow-up (e.g., the high blood pressure and heart disease in radiation-exposed individuals). The men and women working on the Manhattan Project believed that radiation exposure had a "life-shortening" effect.

It is hard to absolutely link long-term, low-dose exposure to ill effects in any single individual since some of these illnesses could occur for other reasons. However, short bursts of high doses of radiation have more predictable dose-dependent symptoms. If the dose is high enough, the individual will die. The first symptoms are usually nausea and vomiting, followed by hair loss and anemia. Many researchers differentiate between two different types of radiation effects:

- *Stochastic effects* include all of the effects of low-dose radiation, including genetic mutations and cancer. There is no threshold for stochastic effects; the probability of damage and disease simply increases with every exposure over an individual's lifetime.
- *Deterministic effects* are determined by a threshold level, above which the effects will occur (see Table 3.1). They generally show up quite quickly. Nausea and vomiting after radiation exposure of 500 mSv is a deterministic effect. The sequence of effects upon a person exposed to 1000 mSv of radiation will include the nausea and vomiting experienced at 500 mSv.

Both low and high doses of radiation increase the probability of stochastic effects, while high doses alone result in deterministic effects.

In summary, radiation dose is cumulative. The probability of genetic defects and various cancers (the stochastic effects) increases over a lifetime, confirmed by their increasing incidence. People undergoing multiple medical tests should be aware of their lifetime exposures. Although exposure for a single dental x-ray is very low, when added to medical and industrial exposure, the subsequent lifetime dose may reach

Table 3.1: Exposure to Radiation

mSv	Health Effect	Time to Onset
50–100	changes in blood chemistry	
500	nausea	hours
555	fatigue	
700	vomiting	
750	hair loss	2–3 weeks
900	diarrhea	
1000	hemorrhage	
4000	possible death	within 2 months
10,000	destruction of intestinal lining, bleeding, and death	1–2 weeks
20,000	damage to central nervous system, loss of consciousness, and death	loss of consciousness within minutes, death within hours to days

Source: Adapted by F. Oelck[11]

the lifetime occupational limit of 400 mSv. Knowing this, an individual might choose alternatives that do not involve radiation, or, in some cases, choose not to have a test at all.

Knowing the health effects of radiation exposure permits health care workers to estimate the amount of radiation to which an individual or group of individuals has been exposed. For example, on March 17, 2011, just days after the Fukushima Daiichi nuclear power plant meltdown, the *New York Times* reported that radiation levels of 400 mSv/hour were being measured at the plant.[12] If a person presented to the emergency room with nausea, vomiting, hair loss, diarrhea, and spontaneous bleeding from the gums or rectum, blood chemistry would be altered and exposure would be calculated to have been at least 1000 mSv. Even if the entire exposure had not happened at once, for example, two hours on day 1, she or he would be exposed to 800 mSv of radiation (400 mSv x two hours = 800 mSv) and a further 200 mSv the next day; the exposures would be additive to the total of 1000 mSv.

Tracking, Limiting, and Preventing Exposure to Radiation

When the radiation source is known, exposure can be limited by controlling for time, distance, and shielding. During the attempts to clean up both the Fukushima and Chernobyl nuclear disasters, protective suits provided shielding for the workers. The most highly contaminated areas were off limits to the workers at first, and the length of time they were allowed to work in these areas was limited (although some took great risks in order to stay on their jobs longer).

As mentioned previously in the book, dosimeters are used to detect and track the levels of radiation exposure for an individual person or a certain area (such as a laboratory or nuclear power plant). The usual dosimeter is a small badge fixed to a person's lapel. Since radiation has no identifying features when it's present—no taste, no smell, no sound, and is invisible—this is the only way that exposure can be tracked. In Canada, a legal framework has been set up to establish the radiation exposure limits for hospital technologists, radiologists, miners, and others working in occupations where exposure is present. Special circumstances such as those involving emergency response teams when the accident involves radiation, for example, a collision involving a truck carrying radioactive materials, also invoke specific regulations and protocols under the Nuclear Safety and Control Act.[13]

Protective clothing should be available anywhere there will be known exposure to radiation. Lead shields or aprons are a familiar sight in dentists' offices. They are used to prevent or limit radiation exposure by covering the parts of the body not involved in the actual x-ray. They are also available in hospital and out-patient x-ray departments. They should be requested if not offered. Personal shielding lowers exposure to radiation by 90 per cent, depending upon the type and thickness of the lead in the apron or shield.[14] Modern x-ray machines, including dental machines, do not produce *scatter*, which are stray x-rays that could affect parts of the body not being examined. Furthermore, any scatter that might be produced will be limited by *radiation shields*, also made of lead, which are installed around the source of radiation.

An extensive number of pharmaceuticals, herbs, and plant products have been studied or are currently under study to mitigate the harmful effects of ionizing radiation.[15] Although some appear promising, the only known useful supplement is iodine, taken as soon as possible after exposure to radioactive iodine. Thyroid-cancer-causing radioactive iodine-131 from a nuclear power plant accident acts exactly like nonradioactive iodine-127 and can potentially build up in the thyroid. Taking normal iodine immediately after exposure fills the sites in the thyroid that use iodine, leaving none available for radioactive iodine. Iodine-127 must be taken as soon as possible after exposure; its usefulness decreases so quickly that by five hours, its protection has decreased by a factor of ten.[16]

Treatment for Exposure to Radiation

The treatment for exposure to radiation begins with *decontamination*. Removing the clothing and shoes from an exposed person gets rid of about 90 per cent of the radiation from an external exposure and vigorously washing the patient's skin also helps. In addition, decontamination also prevents further internal contamination of the patient. (Health care workers wear hazmat [hazardous material] suits while providing care.)

The clinical treatment of radiation exposure depends on the type of exposure, the dose, and the specific signs and symptoms of the victim. Acute health effects from high-dose exposure are usually apparent in three biological systems: the gastrointestinal (digestive tract), blood-making cells in the bone marrow, and the brain and central nervous system. Gastrointestinal symptoms include nausea, vomiting, and diarrhea, which are treated symptomatically with fluid replacement and medications to help the lining of the stomach heal. Bone marrow exposure to radiation can lead to bleeding and/ or infection. Medical treatment includes replacing blood cells, preventing infections, and, in severe cases, replacing bone marrow. White blood cell production can be stimulated by injecting granulocyte-colony-stimulating factor, a hormone used in cancer therapy. Used quickly, it might be shown to decrease deaths by infection. Finally, high-dose radiation exposure to the central nervous system can induce coma, from which some patients have recovered, but usually the result is death; little can be done. Physicians can only provide painkillers and offer support in hopes of recovery or comfort for the dying.

As outlined in the previous paragraph, treatment for radiation exposure entails supporting the body and the systems involved until they have an opportunity to recover, if at all. In other words, it's a short-term, Band-Aid response. Very little can be done about the long-term effects of radiation exposure except by preventing further contamination.

Putting Exposure into Perspective

The greatest contribution to lifetime exposure to ionizing radiation occurs through medical investigations and therapeutic interventions (see Table 3.2). Airline travel exposure due to cosmic radiation varies between 0.003 and 0.0066 mSv/hr.[18] Since all of the activities in Table 3.2 are largely voluntary, they provide a context with which to compare accidental or occupational exposures.

The Linear No-Threshold Model

The *Linear No-Threshold* model (LNT) is the accepted model for the relationship between the amount of radiation exposure and the amount of damage to a biological system. It is used by all nuclear regulatory agencies. It means that the more exposed someone or something is to radiation, the greater

Table 3.2: Average Exposures to Ionizing Radiation through Different Activities

Source of Radiation	Radiation Dose
watching TV	0.01 mSv/yr
air travel (round trip from Washington, DC, to Los Angeles, CA)	0.05 mSv
medical chest x-ray (one film)	0.1 mSv
nuclear medicine thyroid scan	0.14 mSv
bitewing radiographs	0.3 mSv
full-mouth series	0.15 mSv
mammogram (four views)	0.7 mSv
nuclear medicine lung scan	2.0 mSv
nuclear medicine bone scan	4.2 mSv
nuclear cardiac diagnostic test	10.0 mSv
abdominal CT scan	10.0 mSv
various PET studies	14.0 mSv
20 cigarettes/day	53.0 mSv/yr
airport x-ray scan (mostly in US)	0.0001 mSv

Source: Adapted by F. Oelck[17]

the damage. Furthermore, although there is no zero point for radiation exposure (given background radiation), the relationship continues downward to zero. All radiation exposure has some potential effect on biological systems. There is no amount below which there is no harm. Expressed differently, the greater the exposure, the greater becomes the risk of developing cancer, autoimmune disorders, and teratogenic or genetic defects. For example, although everyone is exposed to sunlight, most people don't get skin cancer. On the other hand, the more sunlight to which one is exposed, the higher the risk, a linear relationship. This is further complicated by *shielding*, in the form of skin pigment that can prevent some of the damage. People with less pigment (shielding) are more likely to develop skin cancer.

The LNT model becomes more accurate the higher the dose of absorbed radiation. While high doses kill cells, low doses may leave behind surviving damaged cells that can then multiply and give rise to cancers. The process of healing can repair low-dose damage, minimizing observable effects. Various combinations of these and other effects can occur in a given tissue or organism at low doses. The fact that it becomes harder to specify the effects of low-level radiation, where effects may be characteristically delayed or not immediately apparent, has led to enormous denial, especially among many nuclear industry advocates like Jeremy Whitlock, reactor physicist at Chalk River Laboratories for Atomic Energy of Canada.[20] Head of Health Physics at Gartnavel Royal Hospital in Glasgow, Scotland, Dr. C.J. Martin, points out, "scientific evidence does not show that there is no risk from radiation at low doses. It is at best inconclusive. Political justification cannot be given to a system based on a low dose threshold that could lead to increased exposures with greater associated risk."[21]

Hormesis

Radiation *hormesis* is a discredited "scientific model wherein low level exposure to radiation is not only not harmful but in fact beneficial."[22] It's mentioned here because the theory has persisted in some circles. Hormesis theory postulates that low-level radiation is beneficial to human health until a zero equivalent point (ZEP point) is reached (see Figure 3.2). Above this point, radiation is no longer beneficial and begins to have a negative impact on a person's health. It hypothesized that low-level radiation activates DNA damage repair

Figure 3.2: The LNT Model versus the Hormesis Model

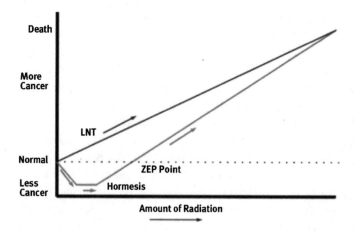

Source: Adapted by F. Oelck [19]

and other protective mechanisms and that the activated mechanisms "overshoot" in their repair so that low-level radiation is actually protective against radiation-induced diseases.[23] Under this model, the body overcompensates in its defence and is better defended in general.

Hormesis does exist—many vitamins and minerals are vital to health in low doses but are toxic in higher doses (e.g,. vitamin A, selenium, copper)—but not with radiation. The theory is still paraded by the occasional supporter of the nuclear industry, like Ann Coulter, a conservative TV commentator in the United States, who advocated it in the wake of the nuclear accident at Fukushima.[24]

Closing Comments

Radioactivity is not good for human beings and other living things. Very small doses are unavoidable because background radiation is everywhere. Damage occurs to molecular processes and structures in cells as a result of the energy transferred from radioactivity. The greater the dose, the greater the biological damage.

CHAPTER 4

Radiation in Medicine

Early Use of Radium[1]

SOON AFTER the discovery of radium by Marie Curie in 1898, travelling snake-oil salesmen were marketing it as beneficial to overall health. The fact that it naturally glowed in the dark made it novel and exciting, and the commercial market took off in 1913.[2] It soon became a popular ingredient in potions and devices intended for curing practically every known or imaginary ailment. It was particularly advertised to treat high blood pressure, goitre, stomach cramps, various women's complaints, kidney problems, and constipation.

Radon was also used for health and beauty purposes, but it quickly fell from grace when consumers discovered they were getting ripped off; with a half-life of about 3.8 days, it was hardly radioactive by the time the product reached the consumer![3] In contrast, radium, with a half-life of 1,600 years, meant that consumers were getting their "money's worth." The US Food and Drug Administration (FDA) initially shut down companies whose products did not contain the level of radiation promised on the label. During the mid-1920s, one high-profile product called Radithor consisted of a micro-gram of dissolved radium in a small, two-and-a-half-inch tall bottle of triple-distilled water. The company claimed "in this bottle there reposes the greatest therapeutic force known to mankind—*radioactivity*."[4] An award of one thousand dollars was offered to anyone who could prove that a bottle failed to deliver at least two microcurie of radioactive radium. No one ever claimed the money.[5]

Radithor was produced by Bailey Radium Laboratories in East Orange, New Jersey. William J. Bailey, the marketer, was a medical school dropout from Harvard University who falsely claimed to be a doctor. Part of Radithor's market-ing strategy was a 17 per cent kickback to physicians who prescribed it to their patients. One such patient was Eben McBurney Byers, who was more famous for his 1906 title as US Amateur Golf Champion than for his title as CEO of Girard Iron Company in Pittsburgh, Pennsylvania. Byers was prescribed Radithor after suffering an arm injury in 1927. Byers soon recovered from his minor muscular injury as might have been expected. Attributing his recovery to Radi-thor, he concluded that if one bottle had healed him, then more would be even better. He ordered Radithor by the case and consumed three small bottles (containing six millisieverts [mSv] of radiation) of radium-laced water daily (giving him a committed dose of 35 mSv per week).

A year later, Byers had severe headaches. His teeth started falling out and his jaw was painful. He was losing weight. His jaw was removed to relieve the pain of the osteonecrosis (a severe form of osteoporosis, in this case perhaps involving multiple tumours as well). Internal radiation ultimately led to Byers's excruciatingly painful death in 1932 at age fifty-one.

After his death, the *Wall Street Journal*, sarcastically referring to Radithor, reported, "The radium water worked fine until his jaw came off."[6]

Radium is an element that decays by alpha emission. As mentioned previously, alpha particles are relatively harmless externally since they cannot penetrate human skin. However, Byers drank them as salts dissolved in water, internalizing them in the process. Radium behaves like calcium, so the body treats it like calcium and deposits it in bones. Since bone doesn't turn over very quickly, the radium atoms tend to stay where they are deposited. Inside the bones, radium damages the DNA of the cells by constantly bombarding them with alpha particles, eventually weakening them. At the same time, radium undergoes radioactive decay to become radon gas, which is also structurally unsound for bones.

The link between Byers's death and Radithor seemed obvious to those around him. He was a prominent man, and the legal suit brought by the radium girls had just been settled. It was clear that the radium water did not live up to its promises. The FDA changed its position on radiation and began closing down companies like Bailey Radium Laboratories. For his part in the Byers tragedy, William J. Bailey was never held legally responsible and died a wealthy man.[7]

During the 1920s, there were other popular radioactive health products making the rounds, such as radium pendants for rheumatism, uranium blankets for arthritis, thorium-laced medicine for digestion, and radium-containing toothpaste, beauty creams, chocolate bars, soap, ear plugs, suppositories, and contraceptives—to name a few![8] In fact, with such a wide market, there were probably many others who suffered the ill effects of radiation before it was pulled from the market.

Although businessmen may have employed dubious market-ing methods by today's standards, their intent was never to cause any damage, only to make a buck. Neither they nor the unwitting public knew what they were dealing with.

X-Rays

As mentioned previously in the book, Wilhelm Conrad Roent-gen, a professor of physics in Bavaria, discovered x-rays in 1895 while he was experimenting with electrons. During an experiment in a darkened room, he covered a cathode ray tube with black paper and noticed that his fluorescent screen was illumi-nated! A projected image of his own hand showed contrast between flesh and bones. He exchanged the screen for photo-graphic film and made the first x-ray, immortalizing his wife's hand by placing it between the tube and the film. He called the new rays "x-rays," after the unknown quantity in mathematics.[9]

Now, for the first time, bones could be examined under their protective layer of muscles and skin. Dentists could identify cavities in teeth, and doctors could find swallowed objects without having to cut the patient open. By 1896, the first x-ray department was set up at the Glasgow Royal Infirm-ary in Scotland and doctors everywhere were quick to add this new tool to their practice of medicine.

X-rays are short, high-energy waves that are released in the form of excess energy when an electron collides with, or has its path altered by, a tungsten target.[10] They are a manufactured form of ionizing radiation; they are not atomic radiation. The process is highly controlled. The technician will set the machine to direct the stream of x-rays to the part of the body being examined. The exact dose and penetration levels are set by a dial. Today's x-ray machines are an extremely sophisticated version of their earlier prototypes.

X-rays are very similar to gamma rays, which also consist purely of energy and have no mass. Like gamma rays, they have the ability to dislodge electrons when they collide with atoms, producing ions and causing oxidative stress. Both the electrons and the highly reactive fragments of molecules (called *free radicals*) that are produced can damage DNA chains and cellular proteins.

Physicians were so enamoured with their new tool that it took practically half a century to understand and accept the degree of damage that could be exacted by x-rays. For example, Dr. Alice Stewart in Britain, in the 1950s, was the first researcher to link x-rays of pregnant women to leukemia in their offspring. The risk was unmistakable: having a chest x-ray during pregnancy increased the incidence of leukemia by a factor of two. Stewart was astounded. She was also ignored—and ridiculed.[11]

The medical community may not have been ready to accept Dr. Stewart's discovery at the time it was made, but then later, in the 1960s, her results were produced independently in the United States by Dr. Rosalie Bertell.[12] It took years for the medical profession to change. Although *pelvimetry*—the taking of three x-rays from different angles to estimate the size of a pregnant woman's pelvis before delivery—eventually fell out of favour, it was still taught in 1976.

Radiopharmaceuticals

The term *radiopharmaceutical* covers a range of drugs, chemicals, or elements that are chemically bound to radioactive atoms, referred to generally as *radioisotopes*. When used for a medical examination, a radiopharmaceutical is usually referred to as a tracer. *Tracers* have a very short half-life (typically a matter of hours) and are introduced into the body by injection, inhalation, or ingestion. They are gamma emitters and, attached to specific carriers, could be considered the ultimate "designer drugs." Wherever they go in the body, they will form an image on photographic film. An example is fluorine-18, a radioisotope that can be produced near the hospital in a cyclotron.[13] Fluorine can be attached to a molecule, such as choline, that will seek out the prostate; simple sugars like glucose, which will be absorbed by actively metabolizing heart muscle; or atoms of sodium to examine bones—each of these will cause the organ to "light up" on film.

Technetium-99m (Tc-99m) is the most frequently used element in radiopharmaceuticals, estimated as 80 per cent of the market.[14] It is produced in research nuclear reactors[15] and transported as its parent element, molybdenum-99 (Mo-99), in a container called a technetium-99m generator (referred to colloquially as a "Molly cow"). The Mo-99 decays to Tc-99m, with a half-life ($t^{1/2}$) of sixty-six hours; the technetium is removed through a chemical process. The *m* in Tc-99m means that it releases pure gamma radiation over a half-life of six hours, long enough for medical examinations yet short enough to avoid major risks to the patient's health. When coming directly from the generator, technetium-99m is attached to an ion specific to the parts of the body being examined. For example, when attached to a simple salt, Tc-99m behaves like potassium in the body, seeking out sites undergoing rapid metabolism so the doctor can identify cancerous tumours or abscesses, or, if attached to albumin, a protein in the blood, the blood flow in the hands or feed can be examined. When a single image is not sufficient, a series of images taken at different times may be more revealing.

Radiopharmaceuticals can also be used to treat disease by virtue of their attraction to, or concentration in, a particular organ or tumour. Because radium behaves like calcium and is absorbed by the bone, it is introduced to the blood stream as a simple salt, radium-chloride (actually radium-dichloride with two chloride ions), to treat bone tumours from metastatic cancer of the breast or prostate.

CT Scans

Computerized (sometimes referred to as "computed") axial tomography, or the *CT scan*, is really a series of x-rays that are computer-manipulated to examine different levels of the body (like slices of bread) or used to form a three-dimensional image of the inside of a person's body. They may be focused to examine only one part of the body as well. The scanner (a sophisticated x-ray machine) looks like a giant doughnut through which the person (or part of the person) lies.

CT scans are extremely useful for examining victims of multiple trauma, such as car accidents, but also in less critical situations like determining the exact extent of a tumour in the lung. Sometimes a *contrast medium*, something that shows up on the x-ray film, will be used and it may be injected, inserted like an enema in the rectum, or ingested depending upon which parts of the body are being examined. The contrast agents are not radioactive; they are substances that are *radio-opaque*, outlining the organs under examination because they produce a barrier to the passage of the x-rays. Historically, barium sulphate, the most common contrast agent for visualizing the gastro-intestinal tract, was, and still is, best for visualizing the inside of stomach walls. However, it is toxic if it leaks into the abdominal cavity, so a sodium salt, diatrizoate, is used more commonly.[16]

The amount of ionizing radiation in CT scans is a concern, as is the ease with which physicians order them. The dose is one hundred to one thousand times greater than conventional x-rays depending upon the body part being examined.[17]

PET Scans

While x-rays show what things look like inside the human body, PET scans show how the body works. As mentioned previously, *PET scan* stands for positron emission tomography scan. The radiopharmaceutical is injected into, swallowed, or inhaled by the patient and picked up by the cells or organ for which it was designed. The PET scanner contains a simple gamma camera that detects the gamma rays and then generates a three-dimensional image in contrasted colours to show how the cells, organ, or organs function. PET scans have largely replaced older gallium-67 scans for cancer diagnosis and staging.

Exposure to a gamma emitter is exposure to ionizing radiation. The average exposure from PET scans is about 14 mSv. The most commonly used radioisotope in PET scans is fluorine-18, with a half-life of 1.83 hours. Other isotopes used in PET scans have half-lives ranging from minutes (122 seconds for oxygen-15) to hours (14.74 hours for yttrium-86).

While PET scans aid in diagnosis, they increase the risk of cancer. In 2011, a team of researchers associated with the

McGill College of Medicine in Montreal reviewed more than eighty thousand patient charts and concluded there was a dose-dependent, increased risk of cancer after the use of these low-dose imaging techniques. For every 10 mSv of radiation, the risk of cancer increased by 3 per cent over the follow-up period of five years.[18]

SPECT Scans

A *SPECT scanner*, or single photon emission computed tomography, uses a moving gamma camera in order to provide a three-dimensional image, and, similar to a PET scanner, involves the injection of a tracer or radiopharmaceutical. The image produced is less localized than a PET scan but is also cheaper to produce. SPECT scans have particular use for examining the brain, in which case technetium-99m is attached to *exametazime*, a molecule that crosses the blood-brain barrier. There is a health risk to using tracers, since no amount of radioactivity is completely without risk. The radiation exposure dose for a SPECT scan is typically in the range of 10 mSv, equivalent to one hundred chest x-rays. Unlike a chest x-ray, however, the patient is radioactive for up to twenty-four hours after the test and exposes friends and family to extremely small doses of radioactivity.[19]

There is concern about the decay products produced from the scans. Even though the tracer itself decays very quickly, its progeny may still be radioactive. The decay product of technetium-99m is technetium-99 (the same element minus the *m*, which represents the gamma ray it has released during decay), which is also radioactive. It decays by beta emission, with a half-life of 211,000 years, to ruthenium-99, which is stable. Once the technetium-99m has decayed, only about 5 per cent of the technetium-99 remains in the body after twenty-four hours, most of it having been excreted in the urine.[20]

Other Scans

Radiopharmaceuticals are used in a number of specialized scans, almost all of which are a variation of PET or SPECT scans. *Sestamibi* or MIBI cardiac scans use technetium bound to a complicated set of six ions (*MIBI* is short for methoxyiso-butylisonitrile). The same compound can be used for examining the parathyroid glands in the neck. In both cases, two scans are required. For the heart, one scan is done at rest and a second after activity to show damage or potential damage. In the case of the parathyroid, the second scan is taken after a "wash-out" period, about two hours later. Most of the radio-active tracer will have gone after the two hours, but abnormal cells will show up on the gamma camera with technetium-99m in the active metabolic "machines," the mitochondria.

MUGA scans (multi-gated acquisition scans) show the heart in motion. Using technetium-99m-pertechnetate, bound to red blood cells, typically sixteen images are taken using the contractions of the heart to trigger ("gate") the pictures. The effective radiation dose is between eight to twelve millisieverts.

Another scan is the *gallium scan*, which uses gallium citrate, one of the earliest and most widely applied radiopharmaceut-icals. The gallium-67 isotope is produced in cyclotrons; its half-life is 3.26 days. The gamma rays emitted during its decay to zinc-65 were picked up on photographic film in early scanners before the advent of gamma cameras and computers.

Radioisotopes with shorter half-lives have overtaken the gallium scan's use in most applications, but the fact that gallium can be absorbed by both dead and alive white blood cells gives it special value in identifying places where they accumulate such as lymphomas, osteomyelitis (infection of the bones), and abscesses. Two scans are usually done, a baseline scan at the time of injection and a second scan two or three days later, allowing the gallium to concentrate in the white blood cells (leucocytes). A short-term disadvantage of gallium scans is that they take up to an hour and a half to complete, during which the patient must remain still. With respect to longer-term health effects, the amount of radiation emitted by the usual dose of gallium-67 will be 25–50 mSv.

Radiotherapy

Ideal *radiotherapy* refers to any method of using ionizing radiation to kill tumour cells without injuring the normal tissues. *External radiotherapy* is generally done by a machine called a linear accelerator, which projects high-energy x-rays at the tumour in order to destroy cancer cells. Treatment rooms are isolated from the rest of the hospital or clinic by shielding. Older radiotherapy units directed gamma rays from cobalt-60 toward the tumour but were limited in the amount of energy available. They continue to be attractive to low-resource settings in the developing world, however, because of their simplicity. *Internal radiotherapy* (also referred to as brachytherapy) is used when the radioactive source can be placed directly into the tumour itself. For example, prostate cancer can be treated by inserting tiny radioactive pellets, iridium-194 being one example, into the tumour.

Of course, radiotherapy has its risks. Although the intention is to avoid irradiation of healthy cells, some will be inevitably exposed to radiation. There is always the possibility that one of these cells will mutate and lead to a secondary cancer. One study estimated that 8 per cent of secondary cancers are caused by radiotherapy.[21] This is difficult to calculate because the same risks present for the first cancer may still exist (for example, smoking) and will vary with amounts of radiation exposure, parts of the body exposed, and the age of the patient.

Closing Comments

Radioactivity, first in the guise of x-rays and then as atomic particles and energy, is widely used in diagnosis and treatment. It never was, and still is not, risk-free. However, most people will exchange the small risk if the benefit is large, such as the diagnosis of a pneumonia. It is difficult to imagine modern medicine without the benefit of ionizing radiation and radioisotopes, but non-ionizing imaging with an ultrasound or magnetic resonance imaging (MRI) is sometimes an alternative. As medical-care costs accelerate, however, physicians and scientists may need to question whether the risk-benefit ratio of costly diagnostic and treatment interventions results in a healthier individual, or even a healthier society. We cannot predict the diagnostic tools of the future. A friend once asked us, "What if, a hundred years from now, doctors regard radioactivity in health care in the same way we now think of bloodletting?"

CHAPTER 5

Industrial Use of Radiation

Improper Disposal of Radioactive Devices: Three Case Studies

T HIS CHAPTER begins with three case studies, all of which illustrate the difficulty of enforcing regulations surrounding the use and disposal of radioactive devices.[1]

Location: Turkey

In 1993, three cobalt-60 radiotherapy sources used for cancer therapy were packaged for transport from Ankara, Turkey, back to the United States. Under international regulations, radioactive devices were to be returned to their country of origin. The cobalt-60 in the device was housed in a steel capsule. Cobalt-60 decays in two steps, through beta and gamma emission, to become stable nickel-60. As long as the source remained in its capsule, no radioactivity escaped. Instead of returning to the US, two out of the three devices found their way to an empty building in Istanbul, where they stayed until the building changed ownership in November 1998. The new owners sold them as scrap metal to two brothers.

The brothers took them to their extended family home and began dismantling the protective containers. Within days, they and people around them became nauseated and started vomiting. (See symptoms of exposure to radiation in Table 3.1.) Doctors initially suspected food poisoning, so the radioactive material remained in the residential area. By the time radiation poisoning was diagnosed, eighteen people had been admitted to the hospital. Ten of these people displayed symptoms of severe radiation poisoning and five of them required hospitalization for forty-five days. Turkish authorities confiscated one of the two sources at a scrap yard before it was melted down. The fate of the second source remains a mystery.

This episode had a remarkably similar precedent: in Brazil, in 1987, scavengers dismantled the lead casing of a radiotherapy device containing cesium-137. The protective capsule broke and powdered cesium poured out. The iridescent blue cesium-137 powder was fascinating; over several days, the thieves, their families, and other people handled it and made designs with it. One child actually ate it! Two people ended up dying, one of whom was the child.[2]

Location: Thailand

In February 2000, a discarded cobalt-60 source was stored on an outdoor lot in Samut Prakarn, Thailand. Two metal collectors bought it and took it to a scrap yard, where the protective covering was cut off with an oxyacetylene torch. Those that came into direct contact with the metal under the covering burned their skin. People close to the capsule felt nauseated practically immediately and some began vomiting later. The symptoms grew worse as the days went on. It took ten days before one of the people affected decided to see a doctor. As before, there was a delay of another seven days before medical authorities suspected radiation was responsible.

A total of about 1,870 people living within one hundred metres of the scrap yard sought medical attention for their exposure to radiation. Ten people had radiation injuries due to direct contact with the capsule; three of these died within sixty days in spite of medical treatment. Thailand's Ministry of Health continues to monitor 258 of the people who lived within fifty metres of the scrap yard for long-term health effects.

Location: France

In November 2000, a man set off a radiation detector on his way into the French nuclear power plant where he worked. Although radiation from the plant was immediately suspected, tests indicated that his body was not contaminated. Instead, to the surprise of all involved, the man's newly purchased wristwatch was contaminated with cobalt-60.

An international inquiry was set in motion. The metal in the watch was traced to a small metal recycling plant in China. That plant had no knowledge of, and did not test for, radioactivity in scrap metal. This particular cobalt-60 probably came from a cobalt-60 gauge or radiotherapy device, was melted down by the recycling plant, and sold, in this case, to a watch factory in Hong Kong. The radioactivity in the watch was small enough to escape detection when it was imported into France, where the worker purchased it. Worldwide investigations did not reveal any additional contaminated watches. However, the small Chinese metal recycling plant had about one hundred kilograms of radioactive metal to melt down and recycle. Without detection and confiscation, this may well have ended up in other consumer products. In addition, the contaminated watch belonging to the nuclear power plant worker was only discovered because he worked where testing for radioactivity was routine; no one knows how much other radioactive waste of this nature has escaped detection. Each item of contaminated material becomes part of the global environment, little by little increasing background radiation.

While these case studies deal with the disposal of devices that may have been used in the field of health care, ionizing radiation has many other applications. Electricity is required to produce x-rays, which limits their portability, while radiation from radioactive elements doesn't depend upon an outside source for its use. Hence, portable measuring devices such as nuclear gauges for industrial or exploration purposes make use of this property. The life-destroying radioactivity can kill bacteria and sterilize selected items. While hardly an exhaustive list, this chapter covers a few common places where ionizing radiation has found a use.

Radioactive Gauges

Radioactive gauges are devices that both emit radioactivity and measure it. The gauge contains a radioactive element. Since it emits ionizing radiation constantly, a shield acts as an on-off switch. The gauge measures in one of two ways: by measuring *backscatter*, the amount of radiation reflected from a material, or, after separating the emitting source and the monitor, by measuring *direct transmission*, the amount of radiation that passes through the substance. Radioactive gauges are used to measure and monitor flow in liquids; to measure and maintain an even thickness of materials such as paper, plastic, or rolled metal; and to evaluate mixtures and compounds for their content, such as determining the amount of water in soil.[3]

Backscatter is measured by placing the gauge on the surface of the material being tested, while direct transmission is measured by drilling a hole into the material and placing the gauge inside. In both cases, a detector counts the radioactivity (gamma rays) reaching it to measure the density. For example, road construction companies make use of backscatter to determine the moisture content and density of asphalt, soil, aggregate, and concrete for quality-control purposes.[4] The radioactive isotopes commonly used in radioactive gauges include cesium-137 (beta and gamma emitter), with a half-life of thirty years, and americium-241 (alpha emitter), with a half-life of 432 years.

When handled properly, radioactive gauges don't pose a threat to human health because the radioactive material in the gauge is surrounded by shielding and the low level of exposure is considered insignificant. The Canadian Nuclear Safety Commission (CNSC) conducts unannounced inspections of industries and companies licensed to handle radioactive materials including gauges. On one such routine visit to a site in September 2013 in Saskatoon, the inspector saw a worker handling a gauge improperly. Not only was the worker using inappropriate technique but he was teaching his incorrect technique to another worker! The inspector ordered the worker to be "immediately removed from activities involving portable nuclear gauges until it can be demonstrated to the satisfaction of the CNSC that this worker has been effectively retrained in all aspects of the safe operation of portable nuclear gauges." A month later, the CNSC was able to confirm that the "corrective measures" were satisfactory.[5] In this case, improper handling of the gauge could have resulted in significant exposure to both workers.

Nuclear Well Logging

Nuclear well logging "includes all techniques that either detect the presence of unstable isotopes, or create such isotopes in the vicinity of a borehole."[6] *Boreholes* are the holes that remain when a core of earth has been removed by a drill in order to prospect for minerals or to examine the subterranean soil structure for heavy construction. Essentially, the process of nuclear well logging is quite simple. A probe (called a *sonde*, with similarities to a Geiger counter) is lowered into the borehole. The amount and kinds of radiation are relayed to a computer on the surface that analyzes the results and interprets them in terms of geological characteristics—porosity, permeability, chemical constituents, and fluid saturation. This information is then plotted graphically.

Industrial Radiography

Industrial radiography involves using penetrating radiation (usually gamma rays or x-rays) to check for internal flaws in welds and castings, and construction flaws in various metal structures. The object being examined is placed between the radiation source on one side and photographic film on the other, like taking an ordinary x-ray. When the object is exposed to gamma radiation, flaws appear as different shades on the film. More radiation passes through any spot that is internally flawed (just as it passes through the fracture site on a broken bone in a person). The most common radioactive elements used in industrial radiography are iridium-192, with a half-life of about seventy-four days, and cobalt-60, with a half-life of almost five years. Both decay by beta and gamma emission.

According to Health Canada, a typical industrial radiography machine produces about two grays (Gy), or two sieverts or 2000 millisieverts (mSv) of radioactivity per minute within a metre.[7] Clearly, the workers using such a device need to be well informed and the location well shielded.

Sterilization

Sterilization refers to the "complete destruction or removal of all forms of contaminating microorganisms from the materials concerned."[8] Sterilization using ionizing radiation is performed in a machine called an *irradiator*, which is essentially a concrete, shielded cell in which the radioactive source is located. Surgical gloves, surgeons' gowns, syringes, bandages, breast and other implants, catheters, and many other health and hospital products are sterilized in irradiators. In North America, about 50 per cent of all disposable medical products are sterilized using radiation.[9] Food products are also irradiated to destroy bacteria; the Canadian Food Inspection Agency has approved irradiation for onions, potatoes, wheat, flour, whole or ground spices, and dehydrated seasonings.[10]

Mail that is sent to important people or places may also be sterilized with radiation.[11]

During sterilization, the object being sterilized is placed upon a conveyer belt that enters the irradiator. It's exposed to gamma rays inside the concrete cell for a specific length of time, transferred to a second conveyer belt, and then leaves, ready for use. The length of exposure varies depending on the standard for the number and type of bacteria left on a given item. For example, implantable devices such as artificial hips or breast prostheses have extremely high standards, much higher than equipment that would be used for colonoscopies. Furthermore, bacteria are more amenable to sterilization by radiation than molds and fungi; viruses are considerably harder to destroy.[12]

Sterilization with radiation has some advantages over other conventional methods. Because of their ability to penetrate matter, gamma rays can reach all parts of the object. Objects can be hermetically sealed and packaged prior to sterilization, meaning that they remain sterile virtually forever until opened. Unlike heat sterilization, radiation has no effect on the temperature of the object, so it's safe to use on heat-sensitive items such as those made of plastic. Compared to gas sterilization, there are fewer chemical reactions and less risk of toxic residue on the object as a result of the process. Finally, radiation sterilization requires fewer external controls. Sterilization via *autoclaving*, which involves heat, or by gas, such as ethylene oxide, requires control of time, temperature, pressure, vacuum-wrapping, humidity, and, in the case of gas, concentration. With gamma radiation sterilization, the only component that requires control is the exposure time.

The most common radioisotopes used in irradiation sterilization are cobalt-60, which has a half-life of about five years, and cesium-137, which has a half-life of thirty years. Both elements decay by gamma and beta radiation. Since sterilization takes place inside a shielded, concrete cell, it's very safe for operators under normal working conditions. In fact, operators aren't required to wear any safety equipment. Products are not even briefly radioactive after the process, and food items that are sterilized by radiation can be consumed safely once they leave the concrete, shielded cell.

Its advantages notwithstanding, irradiation has its own problems. When food poisoning in 2012 struck eighteen people in Canada and more in the United States and sparked the largest recall of beef in history, there were public calls for more irradiation of foods.[13] Yet, it's costly. Irradiators need installation facilities and highly trained staff, not to mention the licensing costs through the CNSC. Plans must be in place to deal with the device when it comes to the end of its operating lifespan. Irradiation is also not 100 per cent effective; it destroys most bacteria that cause food poisoning but could be used to cover up underlying unsanitary conditions in food processing plants.

Another disadvantage is that irradiation of food reduces its nutritional value, especially vitamins. There is certainly controversy over the extent. According to public interest organization Food and Water Watch, irradiation destroys "up to 80 percent of vitamin A in eggs and half the beta carotene in orange juice."[14] Criticism has also been directed toward loss of taste. Ionizing radiation cannot go through objects without affecting them. A trail of chemical transformation—direct impact of the ray or particle upon the cell substance and the secondary transformation when the changed ions react with one another—is left behind. Products found in irradiated food include new hydrocarbons, 2-alkylcyclobutanone from fatty acids, oxidative effects upon cholesterol, and newly produced furans. *Furans* are organic compounds similar to dioxins that are considered toxic and have potential as carcinogens.[15] For all of these reasons, widespread use of food irradiation is unlikely.

Smoke Detectors

The smoke detectors found in most homes and office buildings use americium-241 (Am-241), a by-product of plutonium decay. Plutonium-241, formed in nuclear power plant waste, decays to Am-241, which then becomes a contaminant if the plutonium is being used in mixed oxide (MOX) reactor fuel. Americium is an alpha emitter and a weak gamma emitter. The americium is installed in the ionizing chamber of the smoke detector; alpha particles from the Am-241 knock electrons off the oxygen and nitrogen in the air and turn them into electrically charged ions. These, in turn, allow a very low voltage current of electricity to bridge the gap in the smoke detector. If smoke particles enter the chamber, they attach to the ions and alpha particles and the electricity stops flowing. The alarm then goes off.

Americium is potentially hazardous. Am-241 in smoke detectors is in the form of an insoluble oxide, so that if it were swallowed, it would pass through the digestive system without delivering anything like a significant radioactive dose. If it entered the body in a soluble particulate form, it would concentrate in the skeleton much like radium—with similar health outcomes. When americium smoke detectors were first marketed in 1963, homeowners were required to return them to the CNSC for disposal. Over time, however, regulations changed and now every city dump has a small amount of radioactive waste from discarded smoke detectors.

The alternative to the radioactive smoke detector is the photoelectric cell, in which interrupting the passage of light trips the alarm. The advantage of the radioactive detector is that it can detect particles too small to interrupt the passage of light; on the other hand, it's more prone to false alarms. After claiming the market for more than twenty-five years, the radioactive detector has been receiving harsh criticism. They respond poorly to fire in early stages before flames break out, unless particles are present, and they absolutely fail in high-air-flow situations. Because of these, the International Association of Fire Fighters passed a resolution recommending the use of photoelectric smoke alarms in 2008; this was followed quickly by photoelectric-smoke-detector-only legislation in most Californian jurisdictions.[16]

Airport Security

After the "underwear bomber" of December 25, 2009, the US Transportation Security Administration chose to equip airports and many federal buildings with backscatter x-ray scanners. These delivered a very low dose of radiation, 0.02–0.03 microsieverts per scan, but met with considerable opposition and, in some places, they are already being removed. The European Union banned their use and Canada has not used them. Expectations are that their replacement will be *millimeter wave scanners* that use radio waves.[17]

The metal detectors through which people walk in airports, or the hand-held devices that security personnel use, do not use ionizing radiation. On the other hand, all luggage—unless exempted for specific purposes (for example, camera film)—passes through a cabinet x-ray system. The cabinet is shielded from the people working next to it and the passengers. No radiation lingers on the luggage or any other objects that pass through it.

Closing Comments

Radiation is used to measure and monitor in order to produce a consistent product. It can also aid in determining the presence of water or oil in rock. Ionizing radiation can be used like x-rays to detect faults in construction. There are advantages to its use for sterilization purposes, especially in medical use. None of the radioisotopes used industrially are found in nature; they are all synthetically produced in either cyclotrons or nuclear reactors, although americium-241 is a waste product of nuclear power. Wherever ionizing radiation is used, safety regulations and security measures are involved.

CHAPTER 6

Nuclear Power Plants

The Sarcophagus[1]

U NTIL the Fukushima Daiichi nuclear disaster in Japan in 2011, the worst disaster ever experienced in the civilian nuclear industry occurred at the Chernobyl nuclear plant in April 1986. Initial containment involved hundreds of thousands of "liquidators," men who risked themselves to bring the crisis under control. For intermediate control of emissions, international advisors counselled the Russians to construct a thick, concrete structure to entomb the radioactivity. This tomb was quickly dubbed the "sarcophagus."

In August 1986, four months after the accident, construction began. Improvised lead shields were installed on the drivers' cabins of all the equipment used. Seventy-metre-long beams were transported hundreds of kilometres to Chernobyl. These were assembled like a jigsaw puzzle according to plan. All of the work occurred in a life-threatening atmosphere. Mere minutes of exposure meant the difference between life and death or serious illness.

In September, construction was halted abruptly when the roof of the defunct reactor was too radioactive for work to continue—covered with graphite from the blown-out core, it had to be cleared. Robots were sent in to clear the radioactive rubble, but, within a few days, robot circuits were damaged by the radioactivity, causing them to go berserk and shut down. The bleak consequence was that people had to be used after all. Known as "biorobots," these young army reservists between the ages of twenty and thirty had no right of refusal. In fact, in late August, twelve Estonian soldiers had been executed after they refused to join the clean-up duties. In all, an estimated 700,000 men participated in the clean-up action.

Before beginning their duty, the biorobots sewed their own lead suits, which weighed between twenty-six and thirty kilograms. Major-General Nikolai Tarakanov, who was leading the operation, set a new benchmark for hollow phrases when he stated, in an attempt to reassure the reservists minutes before the operation began: "Comrades, you should know that I was up on the roof two days ago with an Officer . . . and one thing's for sure: I can tell you that there is nothing terrifying."

Going onto the roof in groups of three (two soldiers and one officer), the men had between forty and 180 seconds, depending on prevailing conditions, to throw one or two shovels of graphite over the edge. Then they would run back into the building and the next team would step up. After a "day's work," the men's hands were so weak they couldn't make a fist and they felt as though their "blood had been sucked out of them by a vampire." General Tarakanov would later describe this as "two and a half weeks of hell that lasted 2–3 minutes for each soldier." Each biorobot was given a "liquidators certificate" and a one-hundred-ruble bonus (roughly US$200).

It was thanks to the efforts of the liquidators that construction of the sarcophagus was completed in late November 1986. Estimates were that it would last twenty to thirty years; these proved accurate, as the original sarcophagus is now cracked and leaking radioactivity. Expectations are that the replacement will be completed in 2015. With respect to the liquidators themselves, twenty-eight died of acute radiation sickness and 106 were treated and survived. Their health today is a matter of controversy, the International Atomic Energy Agency (IAEA)[2] reporting the immediate deaths, psychological disorders, and the fact that 350,000 liquidators received a total body radiation of 100 mSv and the Chernobyl Union reporting that 90,000 of the 200,000 surviving liquidators have major long-term health problems.[3]

History of Nuclear Power

On December 20, 1951, the first electricity produced by nuclear power was in Arco, Idaho, where an experimental breeder reactor turned on four light bulbs. The first nuclear power plant that was connected to an electricity grid was at Obinsk, Russia, on June 20, 1954.[4] During the first two decades of nuclear power, there was extensive interest in breeder reactors (see below) for the production of plutonium, however Dwight D. Eisenhower's "Atoms for Peace" speech to the United Nations in 1953 was a plea to use the technology that had produced the atomic bomb for peaceful purposes.[5] The first commercial, electricity-producing reactor in Canada, a CANDU (Canada-Deuterium-Uranium) reactor, began operation in 1971 at Pickering, Ontario (close to Toronto); the first commercial reactor in the United States was the small Yankee Rowe in Massachusetts that operated from 1960 to 1992.[6] France, the country most dependent upon nuclear power for electricity, started commercial production in 1961. Today, thirty-one countries receive some of their electricity from nuclear power, France at 75 per cent to Iran at 0.6 per cent and China at 2.0 per cent.

Nuclear Fission

Nuclear power is generated through nuclear fission. *Fission* refers to the splitting of an atom's nucleus. (The term "fission" was taken from the name for the biological process occurring when one cell splits into two.) The only naturally occurring element capable of splitting spontaneously is uranium-235 (U-235). When it splits, it releases neutrons. The neutrons strike other uranium-235 nuclei, which also split. Inside a nuclear reactor, the neutrons from broken U-235 both start and sustain the reaction (Figure 6.1). Normally, when found in nature, uranium consists of 99.3 per cent U-238 but only 0.7 per cent U-235. The uranium-235 atoms are too far apart for their neutrons to affect one another, so fission chain reactions are virtually impossible.[7] To initiate and sustain a chain reaction in a nuclear power plant, the concentration of U-235 must be increased to 3–5 per cent, a process called *enrichment*.

The actual steps occurring in fission are a little more complicated. Neutrons from uranium-235 collide with other U-235 atoms. When a neutron strikes the nucleus of a U-235 atom, it is absorbed and the U-235 atom becomes an atom of U-236. Uranium-236 atoms are highly unstable and split almost immediately into new atoms. In Figure 6.1, the uranium atom splits into samarian-150 and krypton-84, but the uranium atom could break into many different kinds of atoms, many of them also radioactive. The only requirement in the process of splitting is that the sum total of neutrons and protons in the nuclei of all the products (including the neutrons) adds up to 236.

The U-236 split also results in the release of 2–3 neutrons and large amounts of energy (gamma rays, kinetic energy, etc.). The released neutrons collide with additional U-235 atoms inside the reactor, leading to a *fission chain reaction*. The energy released in this manner is in the form of heat that is used to boil water to produce steam to turn turbines to create electricity. Albert Einstein once famously quipped that nuclear fission was "one hell of a way to boil water."[8]

The Controlled Fission Chain Reaction

To reiterate, a *fission chain reaction* occurs when at least one neutron from a previous fission strikes a new atom to cause another fission reaction. In Figure 6.1, two neutrons are released. In theory, both can be absorbed by subsequent U-235 atoms,

Figure 6.1: Nuclear Fission

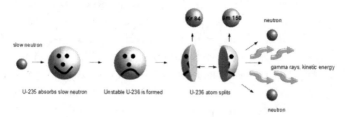

Source: F. Oelck

resulting in four neutrons being released, each hitting another U-235 atom. This could result in eight neutrons being released, and so on. The speed at which this occurs is incredibly fast so if this chain reaction were allowed to proceed unchecked, an explosion would result. In fact, the main difference between a nuclear bomb and a nuclear power plant is the speed at which the fission chain reaction is allowed to proceed. In a nuclear power plant, the fission chain reaction is carefully controlled; in a bomb, it's deliberately promoted.

Nuclear power plants must have a method to absorb neutrons (Figure 6.2). The uranium in nuclear power plants is in the form of pellets stuffed into long hollow tubes called

Figure 6.2: Controlled Fission Chain Reaction

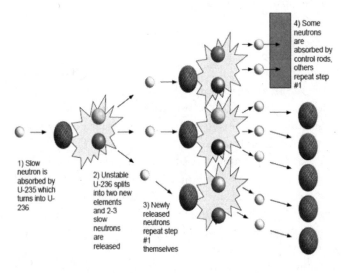

1) Slow neutron is absorbed by U-235 which turns into U-236

2) Unstable U-236 splits into two new elements and 2-3 slow neutrons are released

3) Newly released neutrons repeat step #1 themselves

4) Some neutrons are absorbed by control rods, others repeat step #1

Source: F. Oelck

fuel rods; in order to control the fission reaction, *control rods*, similar rods filled with neutron-absorbing material, are inserted. The further they are inserted, the more neutrons they absorb and the slower the reaction; likewise, the more they are withdrawn, the faster the reaction. Manufacture of control rods is challenging: they must be resistant to swelling during the neutron capture, be able to withstand high temperatures, have reasonable mechanical strength, and survive neutron fluxes without deformation. Two common elements used in the manufacture of control rods are cadmium and boron, but silver, indium, and zirconium diboride (and a host of others including some experimental alloys) can also be incorporated into the stainless steel tubes. Fuel and control rods are combined in structures that are referred to as *fuel assemblies* or *fuel bundles*.

Types of Nuclear Reactors

Nuclear power is used to produce electricity. The energy from the fission reactions—gamma rays, kinetic energy, neutron energy—produces heat to boil water to produce steam. Then the steam is used to turn a turbine, which powers a generator, which produces electricity.

The basic components of a nuclear reactor are the fuel (uranium-235 or plutonium-249), a mechanism to control the speed of the nuclear reaction (control rods or a surrounding liquid or salt), something to heat (water or graphite) that can be recycled, and a cooling system. The most common reactors are pressurized water and boiling water reactors, accounting for roughly 80 per cent of the world's 435.[9] In general, nuclear reactors fall into the following categories:

Pressurized water reactors (Figure 6.3) make up 61 per cent of the fleet of reactors globally.[10] A pressurized water reactor is made up of two separate water circulation systems, a pressurized system in the core and a water-to-steam system transferring the energy to the turbine. In the core, where the fission process is taking place and the heat produced, the high-pressure pump applies enough pressure to keep the hot water in liquid form. The heat from the water is transferred to the second circulation system, turning that water into steam. After turning the turbine, the steam is directed

Figure 6.3: Pressurized Water Reactor

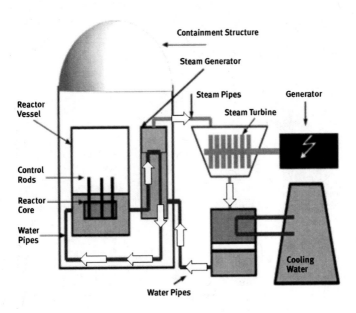

Source: F. Oelck

through a condenser, where it cools to liquid form again and repeats the cycle.

Boiling water reactors (Figure 6.4) make up 21 per cent of global reactors.[11] The key difference is that boiling water reactors have a single system of water circulating through the reactor instead of two. The water boils inside the reactor core where some of it turns to steam. The steam circulates up to the turbine, after which it's liquified in the condenser and the cycle continues. The simplicity of a single system for water is appealing, but the disadvantage is that radioactive water makes the entire system radioactive, not just the sealed system inside the reactor core.

Canadian power plants are all pressurized water reactors, called CANDU reactors, but with one fundamental differ-

Figure 6.4: Boiling Water Reactor

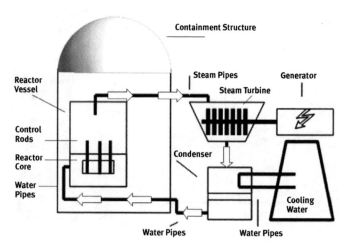

Source: F. Oelck

ence. *CANDU reactors* have heavy water[12] in their reactor cores instead of ordinary *light water* (regular tap water). Water absorbs neutrons that would otherwise be involved in the sustained fission reaction. Because heavy water absorbs far fewer neutrons than ordinary water, uranium doesn't need to be as enriched for CANDU reactors. They can use natural uranium (0.7 per cent U-235) instead of reactor-grade uranium (3–5 per cent U-235). This cuts transportation and cost for enrichment, but this efficiency comes at a different cost: CANDU reactors require more uranium to create the same amount of energy (due to their lower level of U-235). Heavy water is also very expensive (heavy water represents about 20 per cent of the capital costs of a CANDU reactor).[13]

High-temperature, *gas-cooled nuclear reactors* use graphite for modulating neutrons instead of control rods and either carbon dioxide or helium as a coolant. They were used in France and two are still being used in the UK. Their construction was sufficiently complex that, although theoretically capable of producing heat more efficiently than water-cooled reactors, they will likely be phased out.[14]

The Chernobyl nuclear reactor was a *light water graphite reactor* designed in the Soviet Union and was referred to as a RBMK reactor (*reaktor bolshoy moshchnosty kanalny*). The design was principally intended for the dual purpose of producing plutonium and power. To that end, fuel change could occur during operation of the reactor. RBMK reactors use light water as the coolant and graphite as the modulator. Extensive safety modifications have been put in place for the eleven such reactors still in operation in Russia.[15]

Breeder nuclear reactors are capable of generating more fissile material than they consume.[16] As mentioned earlier in the chapter, a plutonium breeder reactor was the first nuclear reactor to turn on light bulbs in 1951, and a thorium molten salt reactor operated from 1965 to 1969 at the US Oak Ridge National Laboratory. They use material like uranium-238 or thorium-232, considered *fertile* materials because, while not capable of sustaining a reaction by themselves, they can be readily changed into another isotope that is fissile. Thorium-232 especially has been touted as the perfect fuel with no need of enrichment, no waste, and no risk of nuclear proliferation.[17] The World Nuclear Association says the "use of thorium as a new primary energy source has been a tantalizing prospect."[18] Thorium becomes uranium-233 after being bombarded by neutrons and the uranium-233 is the actual fuel in the thorium reactor. The neutron source can be external, like a cyclotron, or internal like uranium-235 or plutonium-249. Because breaking an atom into new atoms is unpredictable, uranium-233 isn't the only product. These other broken bits of thorium build up in the reactor and the entire structure eventually clogs up. For example, uranium-232 is an impurity inevitably produced in the bombardment of thorium-232. It's an extremely strong gamma emitter whose decay chain contains other strong gamma emitters. In fact, it is uranium-232 that limits the safety of thorium reactors and makes the manufacture of uranium-233 bombs very hazardous. Yet, in spite of the danger, the United States made a U-233 bomb in 1955[19] and India repeated the feat in 1998.[20] In 2010, the International Panel on Fissile Materials concluded, "After 6 decades and the expenditure of the equivalent of $100 billion the promise of breeder reactors remains largely unfulfilled."[21]

Fusion reactors deserve some mention, even though they remain largely theoretical with one exception: the sun. In the sun, hydrogen atoms fuse together to produce helium and energy. Replicating this process on Earth, some believe, could produce unlimited energy. However, in experiments, conducted at great financial cost,[22] the energy required to make atoms fuse exceeds sustained output energy. For those who dream of "clean" nuclear energy, a fusion reactor is unlikely to be the answer. It will create large amounts of tritium that are difficult to contain and also produce an endless supply of radioactive isotopes, just as current fission reactors do (not likely from the reaction itself but from the surrounding vessels).[23]

Electrical Needs of Nuclear Power Plants

Nuclear power produces about 11 per cent of the world's electricity, representing less than 2 per cent of the world's energy.[24] Electrical supply systems are referred to as *grids*, connecting power lines, transformers, and generators. For an electrical system to work successfully, the power flowing through the grid must be kept at a steady pace with little variation; the network isolates shutdowns so that, for example, a tree falling over a power line doesn't affect the entire grid. Nuclear power plants are both producers and recipients of electricity, requiring electrical power from the grid for start-up, cooling system pumps, and emergency procedures.

Electricity flows through electrical lines in a demand state; it's not stored but rather supplied as it's used. Nuclear power plants, however, need to operate in a steady-state system, called a *baseload mode*, because they respond to changes in demand very slowly. Consequently, they require large industrial or urban customers with consistent demand. The IAEA has other requirements before a nuclear power plant can be licensed: there must be sufficient power elsewhere in the system to supply the nuclear plant when it is shut down for maintenance, ventilation, air conditioning, circulation of coolant, and so on; safeguards for the system must be in place in the event that the nuclear power plant is disconnected unexpectedly; and, for safety reasons, the nuclear power plant itself is required to have two completely separate electrical systems.[25]

Water Needs

Water is used to carry heat from the nuclear reactor core to the steam generator. After the steam has turned the turbine, it's cooled in a condenser, as previously mentioned, but the condenser must transfer the heat to something else. Water is sucked up from a river or lake for this purpose. All steam-cycle power plants (coal, gas, or nuclear) have the same requirement, but it is especially crucial for pressurized water reactors. The IAEA's regulations require them to be sited next to bodies of water; the colder the water, the more efficient the cooling system will be.[26] Most of the water is returned to its source "a few degrees warmer," according to the World Nuclear Association (WNA).[27] This water is not exposed to radioactivity and poses no danger to the environment other than the slight heating. (The cooling pond at Chernobyl was contaminated by the Chernobyl accident, not as a result of normal use.) Yet, problems still occur with water supply. For example, during a heat wave in France in 2007, seventeen reactors had to reduce electrical production or shut down entirely when they were unable to obtain cool water.[28] In the fall of 2013, a Swedish nuclear reactor had to shut down because the cooling water intake pipes were clogged with jellyfish.[29]

Besides requiring water that is cold, nuclear reactors require an enormous amount of it. A typical, one-thousand-megawatt, pressurized water reactor sucks in seventy-six thousand litres (twenty thousand gallons) of water per minute for cooling. This volume of water is equivalent to a room measuring 3 x 3 x 8 m. The water is circulated through an eighty-kilometre maze of pipes and returned at the rate of sixteen thousand litres per minute to the river or stream; the rest turns into steam in the cooling towers. The system is carefully designed so that should a pipe burst, water from the pipe would leak into the power plant rather than run the risk that radioactive water would leak into the environment. As water supply is critical, the industry is seeking strategies to recycle municipal or industrial waste water. As the problem in Sweden illustrated, animal life does get drawn into the intake pipes. The US Environmental Protection Agency estimates that 3.5 billion fish a year are killed by being sucked into nuclear power plant water intake systems.[30]

In addition to an operating supply of water, the IAEA's regulations[31] require a nuclear power plant to have an emergency supply of water, called an *ultimate heat sink* (UHS), which can come from a river or lake or some other dedicated supply. During an accident, the UHS must be capable of delivering 38,000–110,000 litres per minute. For nuclear power plants built in potential drought areas, dams are built to retain a supply of water.

Cooling towers make it possible for nuclear power plants to recycle some of the water. In England, these enormous hyperboloid towers were first used before the 1930s to cool water at a coal-fired electrical power station at Liverpool.[32] The most common cooling tower is a *natural draft tower*, which is composed of a large, concrete structure with air inlets at the bottom, a mechanism to spray the water into the rising air, and a large opening at the top to release the resulting cooled air or steam. The air in the tower can be allowed to rise naturally as it is heated, or forced upward with fans. Forced air towers can be made more compact so they are less conspicuous, but, of course, these require more energy.

Cooling towers consume up to 5 per cent of the water through evaporation, which unfortunately leaves salts and other impurities behind to clog the inside walls, requiring more water for scrubbing. Instead of towers, some reactors in colder climates can simply use open ponds for cooling the water before reusing it, as is done in 14 per cent of the reactors in the United States.[33] While it is theoretically possible to cool down a nuclear reactor without water, with a process called *dry cooling*, where forced air is used much like a car's radiator, it is not currently used in any commercial nuclear power plants. Cooling or storage ponds for used fuel assemblies from nuclear reactors also require water. These Olympic-sized swimming pools continuously circulate the same water, but, as some of it evaporates, replacement is required.

Health Risks of Nuclear Power Plants

The nuclear power industry is a "handle with care" industry. Radioactive exposure requires monitoring and attention to safety. It is probably the most regulated industry in human history. Three specific risks complicate the ordinary operation of a nuclear reactor:

- nuclear meltdown—exposing large numbers of people to an undetermined amount of nuclear radiation,
- low-level planned or unplanned radioactive releases—exposing workers and adjacent populations to low levels of radiation, and
- threat of a terrorist attack.

A *nuclear meltdown* occurs when the reactor core becomes too hot and literally melts. The cooling system in the power plant will have failed. The water will have either vaporized or been discharged. Without constant cooling, the fuel rods melt and containment walls are damaged, and consequently, as in both Chernobyl and Fukushima, enormous amounts of radiation leak into the environment. In the worst-case scenario, the nuclear reaction would be so out of control as to explode like a nuclear bomb.

Fortunately, the risk of meltdowns is relatively rare; there have been only six in the past fifty years,[34] the first of them being in Canada in 1952 at Chalk River, just upstream from the capital city, Ottawa. Accidents involving nuclear power plants are more frequent. The *Guardian* identified thirty-three "serious incidents and accidents" and commented, "information is partially from the International Atomic Energy [Agency]—which, astonishingly, fails to keep a complete historical database—and partially from reports."[35] Wikipedia lists ninety-nine civilian and military accidents from 1952 to 2009 that have "either resulted in the loss of human life or more than US$50,000 of property damage." The total property damage was in the range of US$20.5 billion before Fukushima.[36]

Whether chronic, low-level radiation from nuclear reactors has affected human health has been the subject of many studies. Several studies identified increasing incidences of leukemia in the vicinity of power plants, but study flaws of various kinds made for unconvincing evidence.[37] A German team of researchers, commissioned by the German Nuclear Safety Commission, was formed in 2002 and financed to produce definitive results; the team fully expected that nuclear power plants would be exonerated. The panel included people with a broad spectrum of skills and with varied political positions and financial connections with respect to nuclear power in order to deflect criticism that they could be accused of "cooking the books."

The KiKK (an acronomym for Epidemiologische Studie zu Kinderkrebs in der Umgebung von Kerkraftwerken) study, published in 2007, matched children living at various distances from operating nuclear power plants with those who didn't live close to plants.[38] Matching factors involved things like household environment, smokers in the household, diet, socio-economic status, etc. The researchers concluded there was a positive correlation between residing near a nuclear power plant and the risk of developing childhood leukemia and that the risk decreased the further the residence was from the plant. In other words, the closer a child lived to a nuclear power plant, the greater the risk was that the child would develop leukemia. The authors declined to blame the nuclear power plants and, in fact, stated in their conclusion that they didn't know why the correlation existed.[39] There have been aggressive attempts by the nuclear industry to downplay and discredit the researchers and the results of the study. At a Canadian Nuclear Safety Commission (CNSC) hearing in Ottawa, in September 2010, the audience heard one of the commissioners say that the increase in leukemia was due to a virus![40]

Health experts have been equally quick to point out that nuclear power plants routinely vent radioactive gas when fuel or control rods are replaced. Unless the power plant is recording (and publishing) daily emissions (most do not), the release would simply average out. The levels are usually low enough that they might not affect adults. However, fetuses and children are more sensitive to damage from radioactivity.

The most common radioactive gas that is released by nuclear power plants is tritium. Tritium is a rare isotope of hydrogen (hydrogen-3), formed naturally by the action of cosmic rays from the sun on the upper atmosphere. It is created artificially in nuclear reactors. Like hydrogen, tritium can unite with oxygen to produce water—*tritiated* water. According to the CNSC, tritiated levels in water are most often too low to be detected but seasonal and latitudinal variations may occasionally bring the level up to one becquerel per cubic metre. During the mid 1960s, the tritium from atmospheric bomb testing reached 120 becquerels per litre in rain in Ottawa.

Tritium is worrisome because of its chemical properties. Hydrogen is an essential building block of proteins and enzymes in the body, and the body cannot differentiate between hydrogen that has two neutrons (radioactive tritium), hydrogen with one neutron (deuterium), and hydrogen that has none (normal hydrogen). Tritium can enter the body by inspiration (water vapour), ingestion, or even absorption through the skin. It undergoes beta decay with a half-life of 12.3 years, forming helium-3 (a rare nonradioactive isotope of helium). Although the energy of the released beta particle is quite low, it still has the potential to do damage to biological systems.

In CANDU reactors, tritium is produced when the deuterium (hydrogen-2) in heavy water captures a neutron. The amount of tritium released into the environment from a reactor under normal circumstances is quite low, but sudden spikes are released when fuel rods are changed or the reactor core is breached for any reason. Dr. Ian Fairlie, an independent consultant from the UK on radiation risks, postulates that increases in leukemia around nuclear power plants are associated with these spikes.[41] Although the tritium released is below levels that pose a danger to adults, especially "the reference man," no one has established safety levels for embryos or fetuses. Dr. Fairlie has recommended that plant operators be required to inform the public when potential releases are imminent—pregnant women and children might choose to remain indoors in relative safety.

Finally, the third specific risk posed by nuclear reactors is the threat of terrorist attack, particularly from the air. A 2005 US *Congressional Research Service* (CRS) report stated:

> Nuclear power plants were designed to withstand hurricanes, earthquakes, and other extreme events, but attacks by large airliners loaded with fuel, such as those that crashed into the World Trade Centre and Pentagon, were not contemplated when design requirements were determined. According to a taped interview shown September 10, 2002 on Arab TV station al-Jazeera, Al Qaeda initially planned to include a nuclear plant in its 2001 attack sites.[42]

The United States Nuclear Regulatory Commission (NRC) maintains that it's highly unlikely that a plane crashing into a nuclear reactor could damage the reactor core and cause a widespread release of radioactivity, because "the entire aircraft (including the fuel-bearing wings) is unlikely to penetrate the containment capsule completely."[43] The NRC theorizes that the fuel-bearing wings would need to penetrate the capsule to cause damage. Specific details of security measures around nuclear power plants are not available to the public for obvious reasons.

Nuclear Waste

All nuclear reactors produce large volumes of radioactive waste annually. According to the IAEA, a power plant large enough to serve a city the size of Amsterdam puts out thirty tons of high-level radioactive waste per year.[44] There is enough waste worldwide to cover a football field to the level of seven metres, which may not seem like a very large volume but there is nowhere for it to go. The fact is that the back end of the nuclear fuel cycle has never been given the attention it deserves. Now, decades later, after the investment of hundreds of millions of dollars (mostly public) and dozens of false starts, the waste problem has not been solved.[45]

The waste, the garbage produced by a nuclear power plant, is divided into categories, the *high-level* radioactive waste being the *spent fuel* (*low-level* is all the contaminated metal, clothing, and other things that have had contact with high-level waste). Spent fuel (i.e., the fuel assemblies) is far more radioactive than when it was put into the core. Fuel rods can be handled safely before entering the reactor, being uranium-235 (alpha particle producer), but when they are removed, they contain all the broken bits of uranium, more than two hundred new radioisotopes, producing beta and gamma radiation that would be fatal within minutes to someone standing within a metre from them.

Because the new radioisotopes slow down the production of energy from the fuel rods, nuclear fuel must be removed from a typical, light-water reactor core after four to six years. Since the fuel is highly radioactive and very hot, it is stored on site in *spent fuel pools*, where it is cooled by circulating water. The water acts both to absorb neutrons as the old fuel decays and to cool the used fuel. They are regularly serviced by divers, which illustrates the relative safety of the water as a shield. When the radioactivity (and hence the heat) has decreased to a manageable level (which may take from five to twenty years), the waste is put into dry casks through which an inert gas or air is circulated.[46] Since a permanent location for storing nuclear fuel has yet to be established in any country, and the casks are expensive and awkward to handle, the pools often contain far more radioactive material than originally intended (four times the amount in the case of the United States).[47] The amount of radioactive waste stored in this way concerns the IAEA and other regulatory bodies. An accident involving a spent fuel pool, such as water accidentally discharging from the pool or pool walls being damaged, could precipitate a disaster as extensive as a core meltdown, especially if the pool was packed with recently discharged fuel rods. Nuclear waste will remain radioactive for thousands of years. An ideal, permanent solution for storage would store the waste some place while it decayed safely. The Nuclear Waste Management Organization (NWMO) proposes to store it in deep underground caverns, expected to be safe for 100,000 years, longer than humans have even existed. Computer models, however, indicate the presence of instability and heat production as the radioactive isotopes decay and as radioactive gases are produced. In addition to heat surges, proposals for deep underground storage need to address the potential leakage of radioactivity into aquifers. While no one would plan a storage casket that was not impermeable, the actual effect of continuous radioactivity of this magnitude on the atoms of nickel or iron (steel) over such a lengthy time period is unknown. In fact, the main problem with nuclear power plant waste is that *no one really knows* how it will decay, what new issues might arise, or whether a state of *criticality* (an explosive state) could be reached.

The CCNR has taken the position that nuclear waste should be stored safely on the surface of the earth, close to nuclear power plants. It calls for *rolling stewardship*, a concept first introduced by the US National Research Council in 1995. Nuclear waste would be stored in dry casks, just as it is now. The storage area would be monitored and guarded. Each generation of workers and guards would pass the job on to the next generation. Should something unpredicted occur, like breaching of the casks or sudden surges of heat, the accident would not be hidden underground to wreak havoc. Staff would be able to respond and prevent further damage. While not mentioned by the NWMO, rolling stewardship would be perfect for what it calls *adaptive phased management*, the main difference being that the NWMO proposes putting the waste underground.[48]

On-site storage has another benefit. Every nuclear power plant eventually needs to be *decommissioned*—to be taken apart, recycled, or, in the case of the radioactively contaminated parts, stored. Decommissioning a nuclear power plant is expected to take at least fifty years. The exact time required is uncertain because of differences among reactors and also because no nuclear power plant has yet been fully decommissioned. On-site storage requires establishing a secure perimeter and training staff to monitor waste and adjust management to changing conditions.

The elegance of rolling stewardship includes the enormous financial and environmental savings in transportation costs and solves another problem. Scientists and linguists have puzzled over the problem of identifying the danger of nuclear waste to unwitting generations of humans more than a thousand years into the future. What language or symbols would they use? Under rolling stewardship, the "sign" is in the structure of the stewardship. Furthermore, while today's technology cannot sustainably recycle or neutralize nuclear waste, it is not known what scientists may discover in the future.

Is Nuclear Power "Green"?

Proponents of nuclear power argue that nuclear power is necessary to combat global warming. Even some otherwise astute environmentalists, such as Patrick Moore[49] and George Monbiot,[50] have embraced nuclear power as a solution to the rapidly increasing carbon dioxide in the atmosphere. When operating at full efficiency, nuclear power produces no greenhouse gases. However, there are problems with nuclear power as a "solution" to global warming and the looming energy crisis.

Nuclear power produces only electrical energy, which is a small part of global energy consumption. In 2005, nuclear power produced 16 per cent of the world's electricity; this number declined to 11 per cent by 2012, which is less than a mere 2 per cent of the total energy use of the entire world.[51] Increasing the proportion of electrical power produced by nuclear power plants in order to affect atmospheric carbon dioxide would require a gargantuan global commitment to the industry. It would require an estimated two thousand new nuclear power plants to affect a mere 10 per cent reduction in carbon emissions![52]

The second problem is that the financial cost, the time, and the energy cost involved in building thousands of new nuclear power plants would be staggering. Building and maintaining nuclear power plants is very expensive. Licensing, siting, and the initial construction takes about ten years for each plant. Finally, the input requirements, the energy start-up costs, for nuclear power are large. Carbon-producing petroleum is required for mining; transportation; conversion of fuel; management of tailings and waste (e.g., water circulation in tailings ponds, construction of 30-cm-thick dry casks); and construction and maintenance of the reactors. It takes an estimated twenty years of clean energy production to "repay" the carbon costs of getting started. By all accounts, climate change won't wait.

A third issue is that of waste management and decommissioning. Both of these require energy input. Neither have been successfully completed, so any estimation of energy required is pure speculation. Waste management will require centuries of oversight, and, as mentioned previously, the length of time for decommissioning is estimated at fifty or sixty years. Mine sites also require reclamation. These must be calculated into the total cost of energy required for nuclear power.

Besides failing to be "green," nuclear power plants require an established security infrastructure. An official from the

Tanzanian Atomic Energy Commission (created to explore the possibility of nuclear power for Tanzanian electricity) admitted at a conference in October 2013 that they didn't have the technical expertise, or the system of policies and procedures required by the IAEA, for the nuclear industry.[53] A related problem is also the location of resources; only a few countries (Canada, Kazakhstan, Australia, and the Democratic Republic of Congo, for example) have uranium in large amounts, which means that nuclear-powered nations would be dependent upon them. Would nations tolerate shipments of highly enriched uranium upon the seas? If not, where would more enrichment facilities (risking further proliferation of nuclear weapons) be situated?

Accident at Three Mile Island

The Three Mile Island (TMI) nuclear power plant is located on the Susquehanna River, south of Harrisburg, Pennsylvania, and 257 kilometres (160 miles) upstream from New York City. On March 28, 1979, a secondary cooling circuit in TMI reactor number two malfunctioned. The reactor's primary coolant began to heat up, which caused the reactor to shut down. After the shutdown, a relief valve failed to close properly. Most of the cooling water drained out before the operators were aware of the valve failure because control room instruments showed all valves having closed successfully. About one-third of the fuel rods melted. The reactor vessel itself maintained its integrity and no fuel leaked out. The reactor was designed so that radioactive gas from the reactor was pumped by compressors into waste tanks where the gas could decay. One of the compressors sprang a leak and released between 9.9 and 12.96 curies of radioactive gas, mostly krypton and xenon (according to official reports), during the shutdown. The inquiry into the cause of the leak determined that "deficient control room instrumentation and inadequate emergency response training proved to be root causes of the accident."[54]

The State of Pennsylvania monitored the health of over thirty thousand people who had lived within eight kilometres (five miles) of the reactor for eighteen years from 1979 until 1997 without finding any "unusual health trends."[55] This conclusion is controversial.[56] Critics argue that the studies on cancer rates had methodological flaws, such as focusing on cancers with long latency periods and excluding birth cohorts from the analysis. Authorities claimed that the maximum radiation exposure was one millisievert. Critics point out that this is in stark contrast to reports of people vomiting, losing hair, and developing erythemas (reddened, skin-like sunburns), in addition to the deaths of pets and livestock (see Table 3.1 for radiation levels and corresponding symptoms). As a result, Dr. Steven Wing has argued that the accident at Three Mile Island would result in increased cancer rates at the very time the monitoring stopped.

Chernobyl Today

The sarcophagus built to contain fallout from Chernobyl nuclear reactor number four in 1986 was a remarkable feat in its own right. Even so, shortcuts had to be taken to speed construction and to limit exposure time for workers. The sarcophagus was constructed without welded or bolted joints. Instead, pieces were hooked together or leaned on each other.

Environmental engineers have called it "a house of cards, in danger of collapse after 25 years." In fact, radiation has been leaking through the corroding sarcophagus for years. A collapse would have disastrous consequences because the reactor is still highly radioactive. In order to prevent this, construction of a new cover to replace the sarcophagus, known as the Novarka Project (*novarka* means "new arch"), was negotiated in 2007 and is currently underway.[57]

The project involves construction of two hanger-like structures on rails three hundred metres west of reactor number four. Once complete, they will be joined together and moved over top of the sarcophagus using the rails. The new cover will be 110 metres high (tall enough to fit over the Statue of Liberty), 164 metres wide, span 257 metres across, and weigh more than 29,000 metric tonnes. Robotic cranes and other remotely operable tools will be attached to the inside of the new structure. They will deconstruct the old sarcophagus while specially designed vacuums suck in up to ten metric tonnes of radioactive dust. Most recent estimates indicate that it will be finished in 2015,[58] costing about US$2.08 billion.[59]

Fukushima

On Friday, March 11, 2011, an earthquake with a magnitude of 9.0 struck eastern Japan at 2:46 PM local time, causing reactors numbered one, two, and three of Tokyo Electric Power Company's (Tepco) Fukushima Daiichi nuclear power plant to shut down automatically. Damage to the power plant itself was negligible. However, the earthquake did knock out the plant's own power supply sources. The emergency diesel generators started up. At that moment, the situation was still under control: the fission process within the reactor cores was beginning to slow down and the backup generators were keeping the reactor cool.

However, fifty-one minutes later, a tsunami struck, followed by a second one eight minutes later. The plant's seawater pumps and diesel generators were submerged in water, were damaged, and finally stopped working. Now the situation had changed dramatically. Reactors one and two had lost their power and cooling; reactor three had power for about thirty more hours. The reactor cores one and two began to heat up. They vaporized the cooling water in which they were submerged. Once this had occurred, there was nothing to control the rising temperatures.

That evening, at 7:03 PM, a nuclear emergency was declared, and at 8:50 PM, an evacuation order was issued for people living within two kilometres of the plant. The order was expanded to those living within ten kilometres at 5:44 AM the next day. Within sixteen hours of the earthquake, reactor number one had a meltdown. The heat split the water vapour into hydrogen and oxygen. The gas buildup eventually resulted in an explosion at 3:36 PM on March 12, exactly twenty-four hours after the tsunami. The evacuation zone was enlarged to twenty kilometres.

Over the next couple of days, meltdowns occurred at reactors two and three. On Monday, March 14, a second hydrogen explosion occurred in reactor number three. The next day, on March 15, reactor number four, which up until this point had not been affected, exploded. Hydrogen from reactor three had found its way into number four because of a shared venting system.[60]

Fukushima's Impact on Human Health

Over 160,000 people were evacuated from the area surrounding the nuclear power plant, 100,000 of them during the first twenty-four hours after the emergency was made public. The evacuation process was chaotic as officials tried to determine the extent of the damage. The surrounding area was strewn with debris from the tsunami, and about seventy thousand evacuees were actually directed into the path of the winds carrying radioactivity, the radioactive *plume*. Monitoring and tracking these people is a gargantuan task. People have been traumatized, their social ties shredded, their employment and, often, their self-respect lost, their fears for the health of their children overwhelming, and confidence in the government severely shaken. The magnitude of what they have lost is immeasurable.

In the face of the human side of the disaster, there have been efforts by health care workers to minimize the effects of radiation and treat the effects of psychological trauma. Most of the physical effects will show up only decades after the accident (stochastic effects). The WNA claimed in 2012: "There have been no deaths or cases of radiation sickness from the nuclear accident."[61] On the other hand, the *Guardian* reported that there have been 1,539 deaths attributed to illnesses connected with the accident.[62]

The evacuation, of course, included hospital patients. During the haste, ambulances and similar transport vehicles were overwhelmed and many very ill patients (for example, those who had suffered strokes or end-stage renal failure) did not receive care during the one-hundred-kilometre distance to designated accepting hospitals. More than fifty people died from dehydration, hypothermia, and lack of medical treatment. Of course, radiation did not kill these people. Neither did the earthquake or tsunami. The evacuation order had been given because of fears about the deteriorating conditions at the power plant without the capacity to carry it out properly. Nevertheless, these people count as victims of Fukushima.

While the Japanese evacuated within a twenty-kilometre radius, the NRC increased the zone of safety by recommending that all American citizens within eighty kilometres of the plant evacuate. With regard to children, the Japanese government tried to modify international ionizing radiation exposure standards to allow children to receive twenty millisieverts of radiation per year, prompting a universal outcry.[63] The Fukushima prefecture government plans to conduct lifetime medical check-ups for the estimated 360,000 children who resided in the prefecture during the nuclear accident.[64]

There are enormous challenges to the prefecture's monitoring plan, and, in fact, to any monitoring plan. Can people trust the government to use their health and social information responsibly? The first screening program involved ultrasonography of children's thyroid glands and was fraught with controversy. The government openly declared it was proceeding with the program in order to "calm anxiety" and "convince doubters of the lack of harm." Parents felt disrespected; the results of the screenings were kept from them and they were forbidden, by regulation, from seeking a second opinion. Having been badly conceived, the results of the screening continued to generate controversy. Authorities argued over the number of cysts or nodules normally present in a child's thyroid, the importance of the types of cysts, and whether the plan for recall (only those children with cysts larger than 20 mm or nodules larger than 5.1 mm would be recalled for reassessment) was sufficient.[65]

By August 2013, forty-four thyroid cancers had developed in children in the radiation zone since the accident. Based upon the five-year timeline for thyroid cancers found after Chernobyl, the international press reported that it was too early to have cancers that are linked to Fukushima. However, forty-four cancers in these children exceeds the expected number of cancers in this population by more than a factor of ten.[66] This cannot be a chance occurrence.

Closing Comments

The nuclear industry doesn't set out to have accidents. Each major, and even minor, incident has been followed by extensive reviews of safety protocols at every other nuclear power plant on the globe. The threat of proliferation of nuclear weapons, either by countries or terrorists, will remain as long as there are nuclear power plants. Security services will always be required. The addition to the environment of ionizing radiation by planned or accidental releases of gas or liquids should be a concern for policy makers.

Challenges in international diplomacy exist. Three years after the Fukushima disaster, the accident continues to contaminate the Pacific Ocean while the Japanese government and Tepco have floundered and refused outside help. This refusal raises a number of questions. At what point does a national radioactive crisis, the nature of which expands beyond its borders, require international intervention in the affairs of a sovereign state? At what point should the rights of a corporation to manage its own affairs be superseded by international safety? Since many nuclear reactors require enrichment, and since the same enrichment for reactor fuel can just as easily enrich a bomb, at what point does the international community have the right to curtail the enrichment plans of fellow signatory nations, such as Iran, to the international Treaty on the Non-Proliferation of Nuclear Weapons (commonly known as the Non-Proliferation Treaty)?

Since nuclear power produces waste for which there is no solution, perhaps production should be stopped, at the very least, and a moratorium established on building new plants. If started now, rehabilitating nuclear power plant sites would take the rest of this century to complete, and, as previously noted, require monitoring for generations. While nuclear power plants can be "shut down" in seconds, the nuclear reaction goes on, and on, and on—for tens of thousands of years. There is growing concern among health care professionals about the burden being created for the future. The late John Gofman, a senior figure in the United States Atomic Energy Commission (AEC) in the 1960s and one of the "fathers" of plutonium, resigned when the AEC disputed the health effects of radiation he had discovered. Doctor Gofman stated what many physicians believe: "The nuclear industry is waging a war against humanity. Licensing a nuclear power plant is in my view, licensing random, premeditated murder."[67]

CHAPTER 7

Uranium Mining

The Largest Release of Radioactivity in American History[1]

O N JULY 16, 1979, Robinson Kelly was getting ready for work in the small community of Church Rock in north-eastern New Mexico. That morning, as always, he went to let the horses out of the coral and noticed an awful smell in the air. The little creek on the other side of the yard sounded much louder than usual. Indeed, the former creek had turned into a river, rushing with distinctly yellow-coloured water. There were no signs of rain. Kelly worked at the local uranium mine owned by the Kerr-McGee Corporation. At work, he learned that the Church Rock Dam, holding back tailings, had burst at the uranium mill. Approximately 1100 tonnes of radioactive milling waste and 356 million litres of radio-active waste water had poured into the Puerco River.

The Church Rock Dam disaster is not well known, yet the radiation released there was more than three times that of Three Mile Island (forty-six and thirteen curies,[2] respectively). Uranium mining and mine tailings generally don't garner as much attention by the media, but, in this case, the low public interest and publicity surrounding the disaster was probably due to the fact that the population affected was largely rural, low-income, and Navajo. No research has been done on the health impact of the disaster on the local population. Robinson Kelly probably considers himself lucky to be alive. Both of his parents died of cancer and his uncle, who waded into the water to reach livestock and developed blisters and burns at the time, later succumbed to bone cancer in one foot. Kelly and many others continue to commemorate anniversaries of the spill. Meanwhile, the US Environmental Protection Agency and United Nuclear Corporation (operator of the mill and tailings ponds) continue to argue about clean-up and postpone responsibility.[3]

Mining and Milling

Uranium exists, concentrated in ore bodies, everywhere on Earth. It was present at "the beginning of time" and is believed to be one of the primordial elements in the universe. It becomes profitable for mining at a concentration of 0.10 per cent uranium oxide (U_3O_8) or greater.[4] Until 1954, all of the uranium mined in the world was used for nuclear weapons; that year, the Soviet Union (USSR) put the first electrical-producing nuclear power plant onto an electric grid at Obninsk, about one hundred kilometres southwest of Moscow, followed by the United Kingdom in 1956 at Windscale, Cumberland (now Sellafield, Cumbria), and the United States at Shipping-port, Pennsylvania (about forty kilometres from Pittsburgh) in 1957.[5] Canada, Australia, and the Congo were the largest suppliers of uranium at that time. Today, the top producers of mined uranium are, in order of production, Kazakhstan, Canada, Australia, Niger, Russia, and Namibia.[6]

According to family doctors Linda Harvey and Cathy Vakil, writing for the Environmental Committee of the Ontario Chapter of the College of Family Physicians of Canada, "Uranium mining is the messiest and most contaminating stage of nuclear power generation."[7] The three main ways to mine uranium are by open pit mining, underground mining, and in situ leach mining. Where ore bodies are close to the surface of the earth, *open pit mining* is usual. The surface layer of rock and soil is removed, a gigantic hole or pit is dug, and the uranium is stripped up and trucked out of the cavity. The Ranger mine in northern Australia, operated by Energy Resources of Australia (a subsidiary of Rio Tinto), containing 0.165 per cent uranium oxide, covers a total area of about five hundred hectares (five square kilometres). In 2010, open pit mines produced 25 per cent of total uranium production worldwide (equivalent to 13,541 tonnes).[8]

If the ore body is deeper in the ground, the second method of mining, *underground mining*, requires the building of access shafts and tunnels. The McArthur River mine in northern Saskatchewan is not only the largest underground mine in the world but is also the richest at 16.36 per cent U_3O_8, one hundred times richer than the world's average.[9] Underground mines produced 28 per cent of the world's uranium in 2012 (16,324 tonnes).[10]

When removed from the ground by open pit or underground mining, the chunks of ore are crushed to a fine sand, which is then treated with acid and alkalis to separate the uranium. When uranium is found in unconsolidated material, such as gravel and sand, the uranium is separated while still in the ground by *in situ leach mining*. A mixture of acids, other chemicals, and water—in the United States, sodium bicarbonate, water, and oxygen is the mixture—is forced into the gravel and sand to dissolve the uranium. The resulting liquid is then pumped to the surface where the uranium is recovered by a surface treatment plant. This technology is growing; by 2012, in situ leach mines produced 45 per cent of world's uranium (26,263 tonnes).[11]

In addition to the three main methods of mining, uranium can also be extracted through *by-product recovery*, where the uranium oxide is merely a by-product of mining a totally different material, for example, copper. The only mine doing this in a large scale is the Olympic Dam in Australia, which is the fourth-largest copper mine in the world. Because the steps required to mine copper are the same, by-product recovery occurs for uranium oxide, iron oxide, gold, and silver. The Olympic Dam mine accounted for 6.6 per cent (3851 tonnes)[12] of uranium production in 2012. Since uranium is often found in small quantities in phosphate (for fertilizer) mines, there is interest in developing techniques around extraction from potash.[13]

No matter how it's recovered, from leach solutions (called *leachate*) or acid, or alkali and peroxide treatment, the resulting uranium, in a coarse powder form, is referred to as *yellowcake*. It's packaged in two-hundred-litre barrels and shipped to enrichment facilities. The *tailings*, or refuse that is left behind, whether from ore or leachate, contain about 85 per cent of the radioactivity that was present before the uranium is removed.[14] No longer safely bound in rock, the tailings must be stored carefully in order to prevent dispersal by wind or water.

Storing tailings is a long-term process. In fact, successful and ensured containment has never been done. The usual recourse is to auger the waste into pits that are then covered with water to keep the waste from spreading by the prevailing winds. These are called *tailings ponds*. The theory is that the clay- or granite-lined pond won't leak and neither will the toxic slurry migrate into water systems. Although the World Nuclear Association describes the next step as filling in the pond, covering it successively with impermeable clay, normal soil, and vegetation, it is not clear whether this has actually ever occurred.[15] In any case, there would still be the possibility of radioactive isotopes being brought to the surface by the roots of trees or eventually migrating through the granite lining.

Refining and Enriching

Most uranium found in nature contains roughly 0.7 per cent of the fissionable U-235 and 99.3 per cent of the nonfissionable U-238. After it has been milled into yellowcake, it usually undergoes five more steps before it can be used as nuclear fuel:[16]

1. Refining: The yellowcake is refined to remove impurities. In Canada, this occurs at the Blind River Refinery on the north shore of Lake Huron in Ontario. The resulting, highly refined product is uranium trioxide (UO_3). For use in heavy water reactors (CANDU reactors), it is directly converted into uranium dioxide (UO_2), bypassing the enrichment step.

2. Conversion: For light water reactors, the Port Hope conversion plant, east of Toronto, Ontario, converts the uranium trioxide into uranium hexafluoride (UF_6), nicknamed "hex" in the industry. Hex is highly toxic and requires care in handling; its particular usefulness lies in that it can be a liquid for decanting, a solid for transporting, or a gas for enrichment at very close to normal atmospheric pressures. Since the contribution of fluorine to the mass of the atoms is consistent (fluorine has only one stable weight), the only difference between the atoms of U-238 and U-235 is very small, the weight of three neutrons.

3. Enriching: The majority of nuclear reactors require a concentration of U-235 higher than what is found in most ores to maintain the fission process. Enriching uranium to 3–5 per cent U-235 is done either through gaseous diffusion or centrifuge technology. Both methods take advantage of the three extra neutrons in U-238. With *centrifuge technology*, gaseous uranium hexafluoride is spun very rapidly in powerful centrifuges. Centrifugal force will push the heavier U-238 to the outside of the centrifuge, while the lighter U-235 will tend to concentrate in the centre. The slightly lighter mixture is sucked out of the centre of the centrifuge and put back into a centrifuge. The process is repeated over a thousand times, each time increasing the concentration of the U-235 ever so slightly. *Gaseous diffusion* uses the size difference between U-238 and U-235 to separate them, this time by forcing the atoms

through membranes, thousands of membranes. Being smaller, U-235 will pass through a tiny bit faster, each membrane holding back a few more atoms of U-238. Although centrifuge technology is very energy-intensive, it has become the preferred method, using a mere 2 per cent of the energy required for gaseous diffusion.

4. Second Conversion: All fuel contains uranium in the dioxide form (UO_2). For CANDU reactors, UO_3 is mixed with water, forming a solid, uranium hydrate, which is fed into a kiln to produce UO_2. The UF_6, from the enrichment process, is converted to uranyl fluoride (UO_2F_2), which is then treated in one of two ways. It can be reacted with steam, which removes the fluoride, leaving crystalline UO_2, or it can be reacted with an ammonium compound that removes the fluoride (but leaves the ammonia) and then baked in a kiln to produce powdered UO_2. Although it leaves more waste, the ammonium method is preferred because the powder is easier to manipulate.

5. Fuel Manufacturing: The UO_2 is compressed into pellets, slightly less than one centimetre in diameter and slightly more than one centimetre in length. (A burnable neutron absorber, such as gadolinium, may be mixed into the UO_2 to control the early phase of the U-235 nuclear reaction. As it burns off, more U-235 can react.) The pellets are baked at 1600–1700 degrees Celsius, and then they are packed into fuel rods. The fuel rods are assembled into bundles, which can then be fed into nuclear reactors.

Finally, uranium is also available through dismantling nuclear weapons in a Russian-US program called Megatons to Megawatts. The process is entirely different from that for mined uranium because, in this case, the highly enriched, bomb-grade uranium must be diluted from 90 per cent U-235 to less than 20 per cent. According to the World Nuclear Association, uranium produced as a result of this program has met 13 per cent of the uranium requirements for fuel through to 2013.[17]

Health and Environmental Risks

Miners are continuously exposed to radioactivity. Thorium-230, radium-226, radon-222, lead-210, and polonium-210 are all radioactive by-products of the uranium decay chain and found everywhere uranium is found. Studies of miners working in the 1950s and 1960s have shown a clear link between working in uranium mines and cancer, particularly lung cancer.[18] Although safety requirements for mining have improved enormously since those men were exposed to radioactivity in the mines, the true test of current standards will be the test of time, when today's miners have reached their sixties.

Of the uranium progeny, radon is the biggest culprit with respect to cancer and the greatest challenge with respect to containment. Radon is a gas with a half-life of 3.8 days that decays by alpha emission. In the lungs, alpha radiation bombards the cells in the soft tissue. Radon gas is heavier than air and if it were to escape and be picked up by the wind, it could travel up to a thousand kilometres in four days.[19] Aside from the radon itself, its decay products, radioactive polonium and lead, are highly toxic. Since they are minerals, they would settle out of the air onto crops or water and could potentially enter the human food chain. Because of the movement potential of radon, when it escapes, it would cause contamination well beyond a mine site.

Leaked radioactivity poses health and environmental risks at every step of the refining and enriching processes and during transportation between locations. In fact, no matter what method is chosen to transport radioactive materials, there will always be a risk of environmental contamination and/or injury, as transportation of other hazardous materials has been too clearly illustrated in Lac-Mégantic, Quebec,[20] and not far from Fargo in North Dakota.[21] A small amount of radioactivity is always released during the refining and enrichment processes, and occasionally there are accidents. The chemicals used in the processes are themselves sometimes toxic. For example, fluorides are extremely corrosive, requiring safe storage and/or neutralization.

Finally, any facility that can enrich uranium can enrich it to make nuclear weapons. It takes particular skill to build gas centrifuges and then years of centrifuging to create the 90 per cent U-235 enrichment necessary to make an atomic bomb. Nevertheless, nuclear weapons have been developed under the guise of civil nuclear programs in India, Pakistan, and North Korea.[22] In fact, every country that has developed the bomb has had to do so in secret.

Decommissioning

After a mine site is exhausted, it must be decommissioned and the site returned as closely as possible to its natural state. The

Cluff Lake uranium mine in northern Saskatchewan, operated by Cogema Resources (now Areva Resources) is said to
be successfully decommissioned.[23] Germany, after reunifica-
tion, inherited the then exhausted "Wismut" mine (which
was actually the name of the company who owned it, but popular usage turned it into the name of the entire project). Having started decommissioning the mine in 1991, it has already
cost the German government more than US$9.3 billion and
is not expected to reach completion until 2022.[24]

Decommissioning includes the mines themselves and the
tailings ponds around mining, milling, and processing plants.
In theory, the water in the tailings ponds prevents the radon
gas from escaping. However, water evaporates, concentrating
and exposing radioactive tailings to wind and water erosion,
which could spread radioactive particles into the environment. For example, in 1967, during an extreme drought in
Russia in the Southern Ural Mountains, the tailings pond
at the nearby Mayak nuclear waste storage and reprocessing
facility, referred to as Lake Karachai, did dry up. The wastes
were blown downwind over a broad area about 110 kilometres wide.[25]

Old and abandoned mines litter North America. People
may unwittingly expose themselves to radiation when they
hike or bike through old and often unmarked mine sites.[26]
Radioactive tailings have also made their way into residential neighbourhoods when they have been used accidentally
(or deliberately or unknowingly) for construction purposes.
In Canada, at Port Hope, Ontario, tailings formed part of
the landfill for construction and were used, unbelievably,
beneath St. Mary's, an elementary school.[27] Another example
is construction that occurred in Grand Junction, Colorado.[28]

Unfortunately, uranium mining companies have a dismal record of remediating their mine sites. In northern Saskatchewan, mines closed completely around Uranium City in 1982. Some rudimentary "decommissioning" was performed until 1985, but both the federal and provincial governments agree that more work needs to be done.[29] Watersheds in the area are severely contaminated with uranium and selenium, up to twenty times the limit set out in Saskatchewan's Surface Water Quality Objectives. The Saskatchewan Research Council and Cameco, the latter formed by a merger of Eldorado Nuclear with Saskatchewan Mining Development Corporation in 1988, have undertaken to monitor and clean up the mine sites. Limited funding has been invested in these projects, largely because contamination affects a sparse population in largely inaccessible and poorly resourced northern Aboriginal communities. In fact, Cameco has proposed to mainly "monitor" the sites.[30]

There are uranium mines that have been fully decommissioned. The aforementioned mine site at Cluff Lake in northern Saskatchewan took ten years to decommission; a team continues to monitor the site. Decommissioning is an expensive operation, so new mines must set aside money in trust funds in order to "pay forward" the clean-up costs. There is no guarantee that the costs will be completely covered or that reclamation can or will actually be done.

Returning to the case study offered at the beginning of this chapter, the aftermath of mining radioactive heavy metals has taken its toll on the Navajo nation in the southwestern United States. Not only were miners not informed of the danger of radioactivity during the operation of the mines but they and their families were not informed about the contamination of their water for years. Open mine sites remain, while waste was unmarked and later incorporated into buildings and schools. Governments and companies are stalled in the courts, arguing about their responsibilities while, sadly, nothing is being done.[31]

Closing Comments

Mining is messy, heavy metal mining is messier, and radioactive heavy metal mining is the messiest. Wherever mining, milling, and transporting of radioactive materials occur, constant attention to safety must be maintained.

Mining almost always occurs on Indigenous Peoples' lands, far from urban populations and, certainly in the past, as a result of very poorly planned negotiations between mining companies and governments hundreds of kilometres away in capital cities. In order to access uranium, these colonial powers have often confiscated, if not the land people live upon, then their access to clean water, spiritually significant sites, and traditional livelihoods.[32] Environmental rehabilitation has too often been deferred to the future. If there is any future in uranium mining, due diligence to clean-up must be a priority of mining companies.

CHAPTER 8

Transportation of Radioactive Materials

Nightmare in the Emergency Room

A T 11:00 PM, the doctor is called to the rural hospital. An accident has occurred involving two large transport trucks and a car. The ambulance attendants arrive at the hospital with the first of four victims and inform the emergency room staff that both trucks were marked as transporting dangerous goods. The police are at the scene and two fire departments have been alerted and are on their way. The driver of one of the trucks was pronounced dead at the site.

It was a high-speed accident. As the eight victims, family members, and the assorted onlookers arrive, injuries are documented and the most seriously injured are dealt with first. Two police officers arrive to announce that one of the trucks was carrying two, two-hundred-litre barrels of yellowcake, one of which exploded on impact. A pregnant nurse immediately begs off duty and, in spite of being reassured that the powder would be safe externally, says she will find her replacement. A first responder is directed to the hospital's hazardous materials shower to remove yellow dust from his hair. No one can find the hazmat suits that are supposed to be available for emergencies. The second truck was carrying ammonia.

The emergency department is in chaos. The laboratory technician is searching the Internet for instructions and the hospital's disaster plan loose-leaf binder is on the counter. The most recent copy of Health Canada's *Emergency Response Guidebook*[1] is on the administrator's desk in its original mailing envelope. The police seem to be the most knowledgeable, but, after cordoning off the area, their function is to redirect traffic and wait for the Transportation of Hazardous Materials team to arrive from the capital city two hours away.

This scene is entirely imaginary,[2] but, as Ian Jones, a retired, experienced emergency worker commented after reading it, "It could happen anywhere you have a road and people."[3] Some variation of it has haunted the staff of every small hospital on every highway in Canada.

Packaging Radioactive Material

According to the International Atomic Energy Agency (IAEA), about thirty million packages containing radioactive substances are transported globally every year, including one million within Canada. The majority emit very low levels of radiation (e.g., medical isotopes), so they pose only the slightest health or environmental risk. On the other hand, highly radioactive uranium has been shipped to Chalk River, Ontario, on a regular basis from centrifuges in the United States and, under the US Department of Energy program to repatriate US origin fuels, returned as even more highly radioactive used fuel. Several such shipments have occurred in the last decade.[4] When the University of Toronto research reactor was "decommissioned," the reactor core was sent to the US Department of Energy Savannah River site in South Carolina. Canada's Nuclear Waste Management Organization (NWMO) states that five or ten nuclear fuel shipments are made annually, mostly between operating nuclear power plants.[5]

Packaging regulations for transporting radioactive material are rigid: "All nuclear substances are transported in packages that are selected based on the nature, form and quantity or activity of the substance. There are general design requirements that apply to all package types to ensure that they can be handled safely and easily, secured properly, and are able to withstand routine conditions of transport."[6] Depending on the radioactive substance, packaging falls into one of two categories: certified or noncertified. *Noncertified* packaging material is used for substances that are not highly radioactive, such as medical isotopes, used radioactive gauges, and used packaging. *Certified* packaging is required for substances that are radioactive enough to pose a substantial health risk to the public. For example, nuclear waste and enriched uranium must be certified. Packaging for certification must survive rigorous tests. The package must withstand a *drop test* that simulates helicopter or airplane crashes, remaining intact after falling onto a hard surface from a height of nine metres. It must also survive a *puncture test*, simulated by dropping it a distance of one metre onto a steel rod, and a *heat test*, which involves submitting the package to an eight-hundred-degree-Celsius fire for thirty minutes. The ability to withstand submersion in water is tested by two hypothetical scenarios:

a long immersion in shallow water (eight hours under one metre of water), and a much shorter immersion in deep water (one hour at a depth of two hundred metres).[7]

The same large containers used for dry-cask storage of nuclear fuel waste can be used for transport. These stainless steel cylinders have thirty-centimetre-thick walls, diameters of 2.4 metres, and stand close to six metres tall. Each holds about four tonnes of waste fuel.

Labelling of both the packages and vehicles transporting radioactive substances is also regulated. Specific rules govern the kinds of labels used and where they must be placed. A *Transport Index* indicating the actual dose of radiation at one metre from the surface of the package must be visible on each package. While it is intended to represent the sum of all types of radiation, in practice, it measures only gamma radiation, the beta and alpha being contained by the package itself.[8] The shipping document must include a full description of the nuclear material, a description of the type of vehicle being used, the proposed security measures in place during transport (such as continuous tracking and type of escort), the communications arrangement between the carrier and the response force (i.e., Canadian police agencies), and primary and alternate transportation routes. Finally, *a threat assessment*, a detailed review of people or opportunities whereby the radioactive material might be diverted for unauthorized use, such as in terrorist activities, must be conducted.[9]

Transport Regulations

In every country where transport or shipping of radioactive materials occurs, it is highly regulated. Safety procedures and monitoring processes are largely determined by the IAEA and are modified by different circumstances for each country's purposes. The transportation of radioactive materials in Canada is regulated jointly by the Canadian Nuclear Safety Commission (CNSC) and Transport Canada because there are overlapping responsibilities between them. Transport Canada is responsible for the safe transportation of all dangerous goods. The CNSC is responsible for protecting the environment and Canadian public from radioactive material.

Regulations cover every aspect of transport and require extensive monitoring for compliance. They include certifying packages for nuclear substances; registering users of certified packaging; certifying special and low-dispersible radioactive material; conducting inspections of nuclear shipments for compliance with applicable regulations; licensing carriers, importers, and exporters of nuclear substances (where licences are required); and evaluating radiation protection programs for transport.[10]

Transport

Each person directly involved in the shipment of certified radioactive materials is assigned specific responsibilities. The *consignor* is responsible for preparing the goods for shipping and ensuring that all regulations are met. Radiation levels must be documented for the surface of the package and, as previously mentioned, at one metre away from the surface of the package. The radioactive substances must be stored separately from people and other goods. The *carrier* is the driver of the vehicle or person who is otherwise transporting the goods. The consignor provides the carrier with a shipping document that includes the date, name, and address of the consignor; a detailed description of the goods; the quantity; the number of containers; a twenty-four-hour emergency number where the consignor can be reached; and an emergency response assistance plan that includes a telephone number to activate it. This document must be in paper form and kept in plain view of anyone driving the vehicle, as well as anyone approaching the vehicle from the outside (especially in the event that the driver abandons his post to attend to personal needs or for any other reason).[11]

In spite of rigorous attention to detailed regulations, a regulatory agency is needed to ensure compliance. This role falls to the Canadian Environmental Assessment Agency (CEAA). In Ontario, safety inspections of 102 vehicles carrying radioactive materials during the years 2010 to 2013 resulted in sixteen being taken out of service for violations, such as faulty air brakes, inoperative turn signals, and flat tires.[12] Enforcement actions were taken against nine of the vehicular operators, one of whom had no training whatsoever in dangerous goods transportation, and another who had falsified the transport log.[13]

Risks and Concerns

Secrecy is considered part of security. The CNSC maintains there can be no "public disclosure of security information on the location, routes and timing of shipments, nor any security arrangements or procedures specific to the transport of certain types of packages."[14] The conflict between secrecy and the public's right to know about dangerous threats in their environment is being tested with the announcement in

February 2013 of highly radioactive liquid waste transportation from Chalk River, near Ottawa, to Savannah River in South Carolina.[15] The project is still waiting federal approvals from both sides of the border. There has been massive opposition to the plan. This is the first time that liquid waste has been transferred this distance. In addition to concerns about the level of radioactivity and the amount of traffic (estimated to be about 180 trucks over a four-year period) involved, secrecy about the proposed route(s) was a rallying feature for protesters.[16]

In Canada, accidents have occurred involving transport of radioactive material, but, according to the NWMO, no containers have been breached.[17] This is not an international record, however. A 1996 review based upon Atomic Energy Commission[18] and Department of Energy data counted seventy-two accidents involving nuclear waste transportation ending in the early 1990s. The same report notes, "Descriptions of the events and equipment are insufficient to evaluate the failure mechanisms or sources of contamination."[19] The failure to lodge detailed accident reports leads to an inability to initiate corrective measures and to learn from mistakes.[20]

Physicist Dr. Marvin Resnikoff pursued one such event, occurring in 1980 and described as "surface contamination," and found that it had produced a cascade of subsequent incidents that would have been worthy of a comedy show if they had not been so serious. A cask containing an irradiated fuel assembly accidentally became contaminated during its journey from New Jersey to California, exact details unknown. In order to make it safe for handling, extra lead shielding was added. Unfortunately, in spite of the extra lead, the radiation level in the truck reached double the legal limit. Cask technicians tried to decontaminate it. A capped pipe was opened under observation of a health physics supervisor. Contaminated water poured out of the cask. A technician caught the water in plastic bags and attempted to put the plastic bags into a shielded container. They didn't fit through the opening. He punctured them with a screwdriver to scrunch them and the air was contaminated. Although the water had been measured as being one sievert/hr (1000 mSv/hr) no proper safety equipment had been used. The San Onofre Nuclear Generating Station was fined $125,000 for sending an unqualified supervisor. The subsequent health of the workers is not known.[21]

Carriers are supposed to be aware of their transportation security plan, but, except for spot checks, there doesn't seem to be a process in place to ensure compliance. Recent funding cuts to the CEAA by the Canadian government make it unlikely that the agency will be increasing its monitoring capacity.[22] As the example involving the CNSC and the worker handling a radioactive gauge indicates (Chapter 5), human failure is probably a greater risk than package failure.

In 2005, we contacted the communications or information offices of the Saskatoon Regional Health Authority, the Saskatchewan government, and Cameco Corp. (an international uranium mining company with offices in Saskatoon) for advice on the hypothetical accident described at the beginning of this chapter.[23] All three offered reassurance as their first response. The regional health authority stated that the information would be available in the local facility's disaster plan and the communications manager from Cameco repeated the reassurance a total of three times that yellowcake was safe because the alpha waves couldn't penetrate skin (he dismissed concerns that it could be aerosolized in dust or particulate matter). The provincial government was also reassuring and referred to the *Emergency Response Guidebook* (mentioned in the hypothetical scenario) for further advice. The emphasis on reassurance was disconcerting, especially when it was not accompanied by actual advice.

Closing Comments

The attention given to detailed regulations for packaging and transportation indicates the level of danger that an accident could pose to the health of people and the environment. Regulators report that eighty million tonnes of highly radioactive waste have been transported millions of kilometres worldwide since 1971 without incident.[24] This would be an astonishing record if it were true. Canada boasts some the most stringent and best-monitored regulations in the world. The fact that any infractions of the rules occur at all leads to speculation about what might be happening in less well-regulated environments like India, Pakistan, and the Democratic Republic of the Congo.

CHAPTER 9

Radiation in War

DID nuclear deterrence keep the Cold War "cold"? Does it keep the peace today? Do nuclear weapons really deter war, or has humanity simply played nuclear roulette and "won" thus far?

Nuclear Roulette: Three Case Studies

Cuban Missile Crisis

The world came close to nuclear war during the Cuban Missile Crisis, October 15 to October 28, 1962. The negotiations between President John F. Kennedy of the US and Soviet Premier Nikita Khrushchev were so delicate that they were kept secret; there were many times discussions could have broken down and almost did. The unlikely hero of the time was Vasili Alexandrovich Arkhipov. He was deputy commander of the Soviet *Hotel Class* nuclear-armed ballistic missile submarine *K-19*. The US had discovered that the Russian military were constructing a Russian submarine base on Cuba, so they had created a US naval blockade of the island.

Arkhipov's submarine was ordered to sneak through blockade to assist in the construction. The US navy detected the submarine and began dropping depth charges to force the submarine to surface. The captain of the submarine thought a war between the United States and the Soviet Union had begun. He gave the order to assemble a nuclear torpedo and to prepare to fire on the US destroyer.

Soviet rules for nuclear-armed submarines stipulated that a submarine could use its nuclear weapons only if the three ranking officers on board were in unanimous agreement. The captain and the political officer on board agreed. Arkhipov firmly refused and held his position in spite of enormous pressure from the other two men. Eventually, on October 27, 1962, he convinced the captain to surface. Nuclear conflict was narrowly avoided.[1]

Security Breach in the United States during the Cuban Missile Crisis

On October 25, 1962, a shadowy figure was spotted climbing a fence at the US Army base in Duluth, Minnesota. The sentry who spotted the intruder sounded an alarm that automatically triggered alarms in all of the military bases in the area. Normally this wouldn't have posed a threat since the alarm simply put all the bases on high alert. However, at Volk Field Air Base in Wisconsin, an error in the wiring led to the wrong alarm being triggered. At that base, a klaxon rang out, signalling the beginning of nuclear war.

Practice emergency response drills had been suspended for the duration of the Cuban Missile Crisis, so, as they rushed to their planes and started their engines, the pilots at Volk

Field were convinced this was no drill. This was the action for which they had trained. Planes lined up on the runway for take-off. Fortunately, the base commander called Duluth to find out what was going on and the mistake was discovered. However, the airbase at Volk Field was too small for a control tower, so the base commander had no way of contacting his pilots, who were about to take off. To prevent the launch, an officer from the command post frantically drove his car onto the middle of the runway and began flashing his lights. The pilots stepped down their engines. The suspected Soviet saboteur, the shadowy figure climbing the fence at the base in Duluth, turned out to be a bear.[2]

Russia

On September 23, 1983, Lt. Colonel Stanislav Petrov, a Russian scientist, was asked to replace a co-worker who had called in sick. The job: monitoring Soviet satellites over the United States in order to detect a US first strike. The instructions were simple. He was told to watch computer screens receiving satellite data; if incoming missiles were detected, he was told to push the red button on his desk that would alert Soviet authorities of a US missile launch.

The alarm went off that night as five incoming US intercontinental ballistic missiles (ICBMs) were identified. Protocol demanded that Petrov push the button. However, Petrov took time to think. He reasoned that a first strike would surely involve more than five missiles. He was also concerned that, since the computer system was new, there was a chance it was a mistake. He checked Soviet ground-based radar. It showed no signs of missiles. This was an enormous gamble, since the Soviet Union's time to respond would have been reduced dramatically by the time ground radar picked up the missiles. Petrov opted not to push the button.

Later, a documentary on Petrov entitled *The Red Button* speculated that the person Petrov replaced likely would have reacted quite differently. Regular operators are not scientists; operators follow checklists and rarely make decisions themselves. Scientists investigate. In other words, the normal operator likely would have alerted Soviet authorities that US nuclear missiles were minutes away, sparking nuclear war. The Association of World Citizens awarded Petrov the World Citizens Award in 2006, recognizing his service to humanity.[3]

Details about any recent near misses, taking place in the last twenty years, are hard to come by as most of this information is classified. However, it was not possible to hide the accidental transport of six nuclear-armed cruise missiles from North Dakota to Louisiana on September 5, 2007. Not only were they attached to the wings of a B-52 bomber plane but they were "lost" to military officialdom for over ten hours. At the time, the spokesperson for the US Air Force called it "a departure from our very exacting standards."[4]

Radiation at Sea

In 1954, the United States submarine *USS Nautilus* became the world's first naval vessel to be powered by nuclear energy. The Soviet Union followed suit in 1958 when it launched the submarine *Leninskiy Komsomol*. Nuclear energy greatly lengthened both the amount of time that submarines could remain under water and the distance they could travel. By

2011, roughly 140 ships and submarines were being powered by 180 nuclear reactors worldwide, some of the larger vessels having more than one reactor. The vast majority of these were military vessels, with the exception of icebreakers.[5]

The small nuclear reactors on ships and submarines are similar to the two types producing energy on land (see Figures 6.3 and 6.4 in Chapter 6). Both use U-235, both transport water through various closed circuits, and both use steam to power a turbine. The level of enriched uranium that is used in the reactors varies greatly depending on the country owning the vessel. US submarines tend to use greater than 90 per cent enriched fuel, Russian submarines greater than 20 per cent, French 7.5 per cent, and Chinese 5 per cent U-235.[6]

The fuel's enrichment level affects how often a submarine needs to refuel (refuelling is a complicated process that usually involves cutting open the main hull). A US *Virginia Class* submarine, which uses greater than 90 per cent U-235, can run its entire expected lifespan of thirty-three years without refuelling; the French *Barracuda Class* submarine, which uses 7 per cent U-235, must be refuelled after ten years.

The use of nuclear propulsion results in small nuclear power plants travelling all around the world. There is always the risk that radioactivity will escape into the environment and that fuel will be diverted for weapons production, especially in the case of reactors that use highly enriched uranium. Eight nuclear submarines have sunk since 1963, five from the Soviet or Russian Navies and two from the US Navy. One of the Soviet submarines, referred to as K-27,[7] was deliberately scuttled because it could neither be repaired nor decommissioned at the time. According to the US Navy, the risk of radiation escaping from their two subs is low since they are extremely corrosion-resistant and will remain sealed for hundreds of years.[8] Since nothing has been tested for this length of time, their projections cannot be proven.

Despite the assurances given by the US about its sunken subs, Norway has been sufficiently concerned about the potential contamination of fishing grounds that it has sent teams to investigate the Russian submarine *Komsomolets*,[9] which sank on April 7, 1989, in the Norwegian Sea. When *Komsomolets* sank, it contained one pressurized nuclear reactor and two nuclear torpedoes, each carrying three kilograms of plutonium on board. So far, no contamination has been found. A study published in 1997 by the Norwegian Defence Research Establishment and the Norway Institute for Energy Technology on the environmental impact of *Komsomolets* concluded that "neither the submarine hull nor the reactor vessel will be destroyed by corrosion for at least 1000 years," and that the "effective dose to people whose diet solely consists of fish harvested from the most contaminated waters amounts to no more than approx. 1×10^{-6} of the natural background radiation."[10] This prediction speaks well of the construction of the Russian vessel, but, again, it can neither be confirmed or denied.

Until 1994, naval nuclear fuel and decommissioned nuclear submarines were deliberately deposited in the ocean. Leading up to 1994, the Soviet Union/Russia had dumped more than 148 terabecquerels (TBq) (1 TBq = 10^{12} becquerels) of high-level radioactivity into the Arctic Ocean.[11] (Russia was not alone in using the oceans as a nuclear "toilet"—twelve other countries, including the United States and the United Kingdom, have also done so.[12]) The London Dumping Convention of 1993,[13] followed by the 1996 Protocol, placed a total ban on nuclear disposal at sea.[14] Even so, given the dumping of the past, the possibility of radioactive waste making its way into the human food chain still exists.

While Russia has made some progress in safely dismantling its decommissioned nuclear submarines by bringing them into dry dock, the storage of the radioactive waste is still a problem. Submarines have been dismantled at a faster rate than storage facilities were built to store the waste. Two bases on the Kola Peninsula near Murmansk in northern Russia continue to store nuclear waste in temporary, insecure, above-ground facilities. Concerns about Russia's inadequate storage facilities prompted a financial consortium from several European countries and the European Bank for Reconstruction and Development to support the construction of a long-term storage facility at nearby Sayda Bay.[15]

Russia also relies on ships to transport and store radioactive waste on water. Around seventy ships and barges are used both for nuclear transport and storage in northwestern Russia since no alternative infrastructure (such as rail lines) exists. One of these ships, the former technical support vessel *Lepse*, has been used as a long-term storage ship for radioactive waste since 1986. It saw duty servicing nuclear-powered icebreakers and for dumping nuclear waste at sea from 1963 until 1981. It was moored outside of Murmansk and was on the "nuclear remediation agenda" for decades. On September 14, 2012, many breathed a momentary sigh of relief when it was towed to a shipbuilding yard to be dismantled only to have technical and financial problems plague the project. The *Lepse* remains bobbing in the sea.[16]

One of the conditions of the international Treaty on the Non-Proliferation of Nuclear Weapons (NPT) (discussed in more detail below) is a loophole to give states the opportunity to withdraw fissile material from international safeguards if they intend to use it in nuclear reactors for military propulsion, such as in icebreakers or submarines.[17] This secrecy clause was intended to provide non-nuclear states with the option of exploring nuclear propulsion, but it also means that the International Atomic Energy Agency (IAEA) has no way of knowing if the uranium being removed from its watchful eye is being diverted to make bombs instead.

"Depleted Uranium"

"Depleted uranium" (DU) is a misnomer; it is uranium-238, only considered "depleted" because the U-235 has been removed. It decays by alpha and gamma emission. For every kilogram of U-235 removed for nuclear fuel, roughly seven kilograms of U-238 remain. Otherwise considered waste material, by the 1970s, the US Department of Energy found use for small amounts of DU as ballast in satellites, ships, airplanes, and radiotherapy units. It is extremely dense (an amount the size of a golf ball weighs one kilogram) and difficult to penetrate, leading to research into military use in protective armour and various munitions. As a weapon, it burns on impact and even sharpens as it penetrates its target. As armour for tanks, it is impervious to conventional weapons and can be pierced only by DU ammunition. Depleted uranium armour and weapons were used for the first time in 1991 during the Gulf War. During this short military intervention, a total of three hundred tonnes of DU were scattered throughout Kuwait, southern Iraq, and Saudi Arabia. Subsequently,

DU ammunition has seen action in the Balkans (Bosnia, 1994–1995; Kosovo, 1999), Afghanistan (2001), and Iraq (2003). It was also tested in Japan (1995) and Puerto Rico (1999).[18]

Despite its increased use, employing U-238 in conflict is highly criticized because it burns to produce fine, U-238 oxide dust particles, small enough to inhale. Uranium oxide is 5 per cent soluble in water or body fluids, so while some particles are retained by the lungs, some of the uranium oxide is deposited in lymph nodes, bones, the brain, and testicles. Any targets struck by *DU penetrators*, the name given to U-238, 30-mm bullets, are surrounded by this dust, which eventually dissipates by settling on the ground or being carried on the wind. Thus, in addition to an immediate danger for soldiers in the vicinity of a DU ammunition explosion, the dust also endangers civilians as it travels on the breeze to their homes. Children are at increased risk due to their frequent hand-to-mouth movement, their closeness to the ground, and their increased susceptibility (up to ten times greater) to the carcinogenic effects of radiation than adults.

There has been an increase in cancers and genetic defects in areas where DU has been used. It has been held responsible by many for Gulf War Syndrome.[19] While physicians in Iraq blame U-238 for the effects they have seen, the link between cause and effect is not always clear. It's an understatement to say that war is environmentally unfriendly. In the aftermath of a conflict, there may be other contaminants present with the potential to cause fetal or embryonic abnormalities. For example, the use of Agent Orange (dioxins) in the Vietnam War has shown long-lasting effects on the reproductive health of the Vietnamese.[20]

Nuclear Weapons

To the lay public, a nuclear bomb is a nuclear bomb. To strategists and treaty negotiators, there are two types of nuclear bombs: nonstrategic (formally known as tactical nuclear weapons) and strategic nuclear weapons. *Nonstrategic nuclear weapons* were designed to be used during battle in support of conventional forces. *Strategic nuclear weapons* generally have a much larger destructive force than nonstrategic nuclear weapons and were developed to directly threaten enemy homeland and cities.

The two isotopes used in nuclear weapons are naturally occurring uranium-235 and man-made plutonium-239. In nuclear power plants, uranium-238 absorbs a neutron, becoming unstable U-239, which decays into Pu-239 after a few days. For nuclear bombs, weapons-grade uranium must be enriched to 90 per cent U-235 and weapons-grade plutonium is generally enriched to 93 per cent Pu-239 (with less than 7 per cent of Pu-240).[21] (Weaker bombs can be constructed with lower amounts.) Because light water reactors need to be shut down in order to remove plutonium, IAEA satellite monitors will be quickly alerted when a country is trying to enrich and harvest Pu-239.[22] In heavy water reactors, however, plutonium-239 can be removed while the reactor is in operation, making detection more difficult.

A nuclear explosion can be produced in two basic ways. A fission bomb works by splitting the nuclei of either U-235 or Pu-239 atoms, releasing neutrons and starting an uncontrolled chain reaction. The end result is an enormous explosion. The fusion bomb is the bomb referred to as the "H-bomb," or hydrogen bomb. This is a more complicated process and uses a small fission reaction to provide the force for the fusion and the extra neutrons required. The tritium for the trigger is created when lithium-6 is forced to absorb a neutron. The tritium and deuterium fuse under pressure, forming helium, a free neutron and the enormous energy for the explosion.

The explosion or *detonation* of a nuclear bomb is triggered by either a gun or an implosion method. The *gun method* (Figure 9.1) works only with highly enriched uranium (HEU) and consists of a conventional explosive at the back of the bomb. The conventional explosive launches a *uranium bullet* into a *uranium target*. The two pieces of uranium become a *critical mass*, the amount of a fissile substance required to cause an explosion. As expected, an enormous explosion results. The gun-assembly nuclear weapon is the easiest nuclear weapon to construct but requires about fifty kilograms of HEU, a fairly large amount.

Figure 9.1: Gun-Assembly Nuclear Weapon

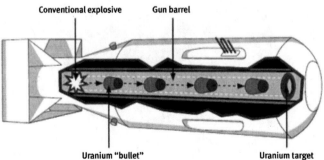

Source: International Commission on Nuclear Non-proliferation and Disarmament (ICNND)[23]

In the *implosion method* of detonation (Figure 9.2), uranium-235 or plutonium-239 is bunched into a core and surrounded by conventional explosives. When the explosives are detonated, the material in the core compresses, reaches critical mass, and undergoes a chain reaction followed by a nuclear explosion. The implosion-type nuclear weapon is far more complicated to construct but requires much less fissile material—just five kilograms of plutonium-239 or two kilograms of uranium-235. With the use of *neutron reflectors*—materials such as beryllium, graphite, or steel surrounding the explosive material that reflect neutrons back to it and improve detonation devices—both types of nuclear weapon have become easier to construct.

Investment in nuclear weapons occurs because of the

Figure 9.2: Implosion-Type Nuclear Weapon

Fast explosive Slow explosive Tamper/Pusher

Neutron initiator Plutonium core Spherical shockwave compress core

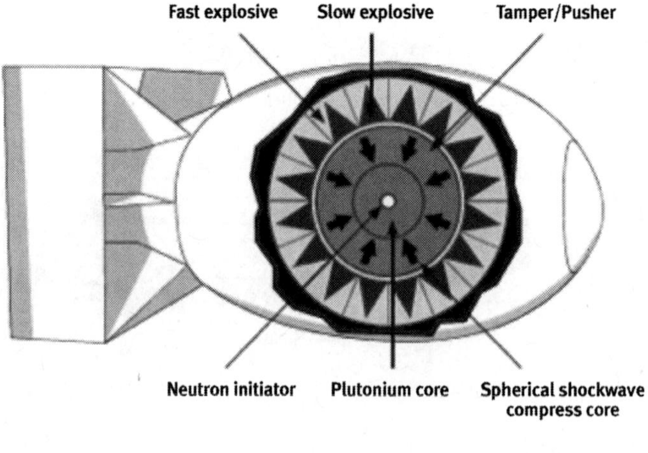

Source: ICNND [24]

perceived inferiority of conventional forces among hostile states (e.g., Israel) or because of perceived threats from superpowers or neighbours (e.g., North Korea, Pakistan). The investment comes at a great cost to health, social programs, and infrastructure.[25] The estimated cost of the US nuclear program for the next decade is around $620 billion.[26] All of the countries with nuclear weapons are listed in Table 9.1, the numbers being the result of self-reporting, IAEA inspections, and, in the cases of Israel, Pakistan, India, and North Korea, estimations based on undercover investigations. Transparency is not generally part of nuclear weapons

Table 9.1: Global Inventory of Nuclear Weapons[27]

Country	Total Inventory
Russia	8,500
United States	7,700
France	300
China	250
United Kingdom	225
Israel	80
Pakistan	100–120
India	90–100
North Korea	‹ 10
Total	~17,300

programs, so there are discrepancies in the numbers provided. Thus, the table includes rounded numbers and is based on incomplete information and expert estimates.

The size or *yield*—explosive energy output—of a nuclear bomb is compared to the explosive force of *TNT* (the conventional explosive used in sticks of dynamite or bombs, technically called trinitrotoluene) and is measured in tonnes (t), kilotonnes (Kt), and megatonnes (Mt). The nuclear bomb "Little Boy," dropped on Hiroshima, Japan, in 1945, had a yield of 15–16 Kt, equivalent to 15,000–16,000 kg of TNT.[28] The Soviet Union constructed and detonated the most powerful nuclear weapon, dubbed "Big Ivan," in 1961. It had a yield of fifty-seven megatonnes, produced shock waves around the world, and blew out windows nine hundred kilometres away in Norway.[29] Today's bombs are in the range of 100 Kt (over six times the force of the Hiroshima bomb) to 550 Kt.

Although the technology to produce a nuclear bomb has become simpler, it is a different type of nuclear weapon that is more likely to be used by a terrorist group. So-called *dirty bombs* contain mixtures of a conventional explosive and concentrated radioactive material. The explosion disperses the radioactive material into the surrounding area. This would not require the sophisticated technical expertise that a nuclear weapon requires nor would it require a fissile substance. A conventional bomb containing cobalt-60 or cesium-137 would be sufficient to make a large area uninhabitable. With respect to fissile materials, in spite of stringent regulations about the transport of radioactive materials and painstaking tracking of fissile materials, the IAEA reports that nuclear and radioactive materials are still going missing from the world's inventory.[30] There is enough missing fissile material to make dozens of nuclear bombs.

A discussion about nuclear bombs wouldn't be complete if the theoretical "doomsday bomb" were not included. The "doomsday bomb," first proposed in 1950, is actually a specific form of a "dirty bomb" that creates its own cobalt-60 with the neutrons that are used to start the explosion. The Co-60, with very strong gamma rays and a half-life of 5.24 years, becomes airborne with the explosion, eventually spreading widely. The number of dead would overwhelm the few who survived. People would be dead, but physical structures—houses, cars, jewellery, valuables—would remain and be accessible after several half-lives of Co-60.

Nuclear War[31]

Nuclear war—or any exchange of nuclear weapons anywhere on the planet—would have both local and global consequences. There would be an immediate impact on the city where the bomb is detonated and a longer-term environmental impact, both locally and globally. The blast of the bomb would create a shockwave of air pressure outward from the explosion, expressed as an extremely strong wind. The force would be enough to flatten brick buildings, crush human lungs, and turn humans and objects into missiles flying at several hundred kilometres per hour.

In the explosion of an average bomb today of one hundred kilotonnes, intense heat and light would initially be released. Close to the blast site (*epicentre*), nearly all materials would immediately vaporize. A gigantic fireball would expand outward about fourteen kilometres, consuming all available oxygen in the process and producing a pressure wave referred to as *blast pressure*. The blast pressure would generate winds strong enough to blow out windows up to twenty kilometres away from the epicentre. Direct exposure to alpha, beta, and gamma radiation could kill people quickly (close to the epicentre) or in the long term (from various cancers), depending on the intensity, type of radiation, and type of exposure. Those "lucky enough" to avoid the blast pressure and blindness by the intense light (stretching up to eighty kilometres away), would suffer burns on their skin, or be injured by the ignition of flammable material (flammable material stored kilometres away from the epicentre can be ignited), or be injured by flying debris. Electromagnetic pulses would disable all kinds of electronics, such as microchips in cars, phones, computers, and medical equipment, and, although not a direct threat to human health, this would undoubtedly lead to further casualties. The havoc and panic wreaked by the loss of communications technology would be enormous.

Fallout from the blast could carry the danger far beyond the epicentre. *Fallout* is the debris that falls back to Earth after the bomb blast. It consists of aerosolized materials from the site of the blast that have been irradiated and propelled upward by the force of the blast. This debris makes up most of the mushroom cloud typically associated with nuclear explosions. The spread of the fallout would be influenced by the weather and topography of the land. Wide swaths of countryside could be contaminated. Given that some radioactive elements have half-lives of thousands and even billions of years, the radioactive legacy of a bomb could last essentially forever.

The impact of a nuclear bomb is dependent on many factors: explosive yield; type of fissionable material; location of detonation, whether high above the target or burrowed into the earth; weather; topography of the land; and more. Just as it would be impossible to predict the exact impact of a nuclear bomb, it is impossible to create a definitive, planned response to a nuclear attack. The British Medical Association estimated that 90 per cent of the physicians in Hiroshima were direct casualties of the bomb, as were a similar proportion of nurses and ancillary health care workers.[32] It is likely this would occur in the event of another war. As in Hiroshima, surviving doctors would be overwhelmed with the sick, injured, dying, and dead. There would be no effective medical response.

Wherever one bomb was dropped, or wherever there was an exchange of weapons, the resulting death toll would be enormous. Emergency response teams from around the world would be sorely taxed to respond. In addition to the usual disaster response, those that could assist would be severely hindered by radiation, fallout, fires, general panic, and loss of critical infrastructure.

Nuclear Winter[33]

Nuclear winter describes the effects that a nuclear war would have on Earth's climate. Scientists have calculated that even a small-scale nuclear war involving one hundred Hiroshima-type (fifteen kilotonnes) nuclear bombs between two countries such as India and Pakistan would have a devastating effect on Earth's climate. Such a war would create approximately five megatonnes of black smoke, which would rise into the stratosphere from burning cities. At stratospheric levels—

most commercial aircraft fly in the lower stratosphere at 9144 metres—it would spread around the world, blocking out sunlight. The ozone layer would be destroyed, which would hinder eventual recovery because once the atmosphere cleared of debris, anything surviving would be exposed to ionizing radiation from the sun. There would be no rain in the stratosphere to wash the smoke out, so it would last for years. By blocking the sun, the limited nuclear war would also plunge the planet into temperatures too cold to support agriculture. The drop in temperature would be practically immediate. Any exchange of more, or larger, weapons would probably create climate changes that would last decades.[34]

The impacts of this small-scale nuclear war would cascade onto just about every system of life on the entire globe. The smoke and radioactive fallout would include the toxic gases produced when cities, and their contents, burned—synthetic materials, cyanides, dioxins, and furans, to name just a few. It is unlikely there would be any survivors, but should there be any, as "winter" thaws and the sun returns, they would be dealing with billions of thawing animal and human bodies.

Finally, one last fact: this hypothetical, small-scale war between India and Pakistan described above would, at most, involve only 0.3 per cent of the world's nuclear explosive power.[35]

Nuclear Disarmament

The threat or use of nuclear weapons violates international humanitarian law. The law of armed conflict prohibits unnecessary slaughter of noncombatants and long-term, widespread, or severe damage to the environment. It also prohibits damage to neutral states.[36] Nuclear weapons would contravene all three of these provisions. In 1996, the International Court of Justice concluded that since international humanitarian law (IHL) did apply to the use or threat of use of nuclear weapons, their use would be a violation of IHL principles.[37] In 2011, citing IHL, the International Red Cross/Red Crescent called upon "all States . . . to pursue a complete ban on nuclear weapons" and in 2013 adopted a four-point plan toward the prohibition and elimination of nuclear weapons.[38]

Earlier efforts to contain and roll back the spread of nuclear weapons resulted in nations of the world coming together in 1968 to produce the Treaty on the Non-Proliferation of Nuclear Weapons (NPT). The NPT has three intentions: disarmament by nuclear weapons states; prevention of new nuclear weapons states; and promotion of nuclear power. The NPT came into force in 1970. Over 190 sovereign states have signed the treaty—more countries have ratified it than any other arms limitation and disarmament agreement. Every five years, with the last meeting occurring in 2010, countries convene at the United Nations (UN) headquarters in New York City to work on it. In 1995, the term of the NPT was extended indefinitely. In return for renouncing nuclear weapons, signatories are allowed to develop nuclear power for domestic energy.[39]

While laudable in theory, in practice, many of the NPT's own signatories refuse to abide by its stipulations. This includes the stubborn refusal of some to disarm their nuclear weapons. In fact, flouting the treaty, the United States is "modernizing" its nuclear weapons.[40] There are also countries outside of the NPT that possess nuclear weapons—for example, India, Pakistan, Israel, and North Korea. Meanwhile, the IAEA finds itself in a conflict of interest; charged with policing weapons production, it is also in the business of promoting nuclear power. Hence, relatively little criticism has been levelled at Canada, a signatory to the NPT, for trading in nuclear technology and uranium with India, a nonsignatory, even though this is forbidden under treaty protocol.

In 1998, frustrated by the failure of the NPT to result in a clear and defined timeline for destruction of nuclear weapons stockpiles and the unbalanced bargaining table, where any one of the five permanent members of the UN Security Council can veto decisions made by the entire body of the UN,[41] the Middle Powers Initiative (MPI) was created. Led by then Canadian Senator Douglas Roche, this collaboration of eight international nongovernmental organizations (NGOs)[42] works primarily to support middle power countries in their efforts to seek abolition of nuclear weapons. *Middle power countries*, by their own definition, are "politically and economically significant, internationally respected countries that have renounced the nuclear arms race."[43] Significant members of the MPI include Norway, Brazil, Canada, Australia, and Argentina. One of the MPI's notable initiatives has been the Article VI Forum that focuses on the political, legal, and technical elements required for a nuclear-weapons-free world.

Parallel to the MPI is the international organization, Parliamentarians for Nuclear Non-Proliferation and Disarmament, a forum for like-minded elected members of governments to share plans and strategies. In another effort to disarm globally, geographic regions have joined together to create Nuclear Weapons Free Zones. The complicated negotiations between states to reach such agreements serves to strengthen resolve for nonproliferation and abolition of nuclear weapons.

One of the MPI members, Costa Rica, frustrated by Security Council roadblocks, in 1997, on behalf of other states' legal, scientific, and disarmament experts, submitted a model Nuclear Weapons Convention to the UN Secretary-General. Much like the Land Mines Treaty that was reached through the Ottawa process, bypassing the UN,[44] a Nuclear Weapons Convention would bypass the NPT. The International Campaign to Abolish Nuclear Weapons (ICAN) relaunched the concept of a Nuclear Weapons Convention in 2007. It was later promoted by UN Secretary-General Ban Ki-moon in 2008. A global poll showed 76 per cent of the public supported such a convention.[45] In Canada, with the help of Canadians for a Nuclear Weapons Convention, both the Senate and the House of Commons passed motions unanimously encouraging the Government of Canada to "join negotiations for a nuclear weapons convention."[46]

Closing Comments

Radioactivity is used in conflict; it's harnessed to propel ocean-going vessels and finds its way into weapons in the form of uranium-238. Depleted uranium is an illegal weapon, given that it pollutes Earth long after the conflict is over.

Atomic weapons raise the amount of radioactivity released in war by logarithmic amounts. They are valueless as weapons against terrorists or against internal threats because of their indiscriminate lethal effects. The tenacity with which nuclear-armed states hold onto their weapons, even in the face of rising maintenance costs, strongly suggests their symbolic value is great. Since there are close to twenty thousand nuclear weapons in existence, the threat of nuclear war by accident or intention exists. Unless nuclear-armed states begin to live up to their obligation to disarm under the NPT, other countries will want nuclear weapons as well. The more countries that have nuclear weapons, the more likely they will be used, either by intention or by accident.

International negotiations led to the abolition of slavery and have been instrumental in decreasing racism globally. When nuclear weapons become sufficiently morally reprehensible and the desire for the abolition of nuclear weapons outweighs the psychology supporting them, they will enter the dustbin of history.

CHAPTER 10

Conclusion

T HIS BOOK has aimed to provide its targeted audience, readers without a scientific background, with enough information to understand radiation in general, the impacts of events like the Fukushima meltdown, and the extent of responsibility ionizing radiation places upon the future. To this end, we have tried to answer relatively straightforward questions: What is radiation? How has it been understood or misunderstood over time? Where do we encounter it? How does humanity use it? What are the corresponding risks and benefits?

The twentieth century marked the beginning of the nuclear age. People learned to use ionizing radiation to diagnose disease and treat patients, provide energy to heat and illuminate homes, kill dangerous bacteria and germs on medical equipment, and delve into the earth for dense materials. Researchers will continue to explore the science of ionizing radiation and atomic energy.

Yet, with all of its potential, ionizing radiation has also been responsible for many deaths, possibly even millions of deaths from years of silver mining in the presence of "bad luck rock," plus the deaths resulting from x-rays and atomic bombs and those from natural radon in homes. The questions we ask are: How much risk is unavoidable? What risks are we able to modify? When do the risks outweigh the benefits? The risk of a nuclear meltdown is extremely small, but the effects of the ensuing social upheaval, environmental contamination, and risks to human health are enormous, as has been seen in Fukushima, Japan. Fukushima put the question of nuclear energy back on centre stage, leading some countries like Japan and Germany to decide that the benefits of nuclear energy no longer justify the risks. Meanwhile, other countries such as India and China have reaffirmed their commitment to nuclear energy.

Ionizing radioactivity can be both a friend and foe. It fascinates us, much like the phenomena of black holes and deep space, and like them, our understanding of it remains somewhat limited. It is an enormous force that we cannot tame. The nuclear age is determining the level of background radioactivity for future generations; each time we add a little, we inevitably and unavoidably affect the future. Through the nuclear industry, we are putting the burden of waste management on future generations, incurring an unknown debt to the future.

Our current practice is unsustainable. We cannot continue to render parts of the world uninhabitable by plan (mining activities and uranium-238 weapons) or by accident (Chernobyl and Fukushima). In fact, the contaminated areas contaminate the whole. Radioisotopes from Fukushima have been found in California and Lithuania, while Chernobyl fallout extended to Cincinnati and Finland.[1] In 2006, at the International Physicians for Prevention of Nuclear War (IPPNW) congress in Helsinki, Finland, Punjabi doctors revealed their concern that uranium-238 from the Gulf War had contaminated their valleys.[2] There is absolutely no doubt that this generation is making the world more radioactive.

Proponents of nuclear energy continually compare accidental or planned releases of radioactive materials to background radiation in order to reassure the public of the safety of the release.[3] The problem with this reassurance is that there is no safe level of radiation. Each amount increases the risk of cancer and once background radiation is increased, we don't have the means to reduce it. It is safe to say that no one fully understands ionizing radiation; certainly it has not been "tamed" as once hoped. We cannot contain it safely and completely.

Ionizing radiation seems to have a place in our industrial and health care future. Through market values and enforced governmental regulation, we have a chance to determine its usefulness and what risks we are willing to take to justify that use. In order to make decisions about nuclear power, nuclear weapons, tests using ionizing radiation in health care, choices about nuclear waste sites, safety in transportation, and the uses of any radioactive devices, everybody needs to be informed. Without the public knowing the trade-offs, nuclear power companies are able to politically divide communities and governments, take advantage of lower socio-economic and less well-educated communities, and spread glossy propaganda in schools.

Whether people are pro-nuclear or anti-nuclear, ionizing radiation affects us all. There is work to be done. Existing nuclear power plants must continue to maintain and heighten their safety and security systems. They must develop decommissioning plans, monitor waste, and remediate mine sites.

We, and the industry, cannot deny responsibility for the health and welfare of our descendants. Nuclear-armed nation states must dismantle their nuclear weapons. Individuals must be empowered to make choices based on knowledge, not fear. Collaboration is essential. Our challenge is to figure out how to live with ionizing radiation and deal with it safely today—and for a very long time in the future.

Further Reading and Online Resources

W E WOULD RECOMMEND the following books as further reading on ionizing radiation, its history, and its effects on the health of individuals and the environment:

Bertell, Rosalie. *No Immediate Danger: Prognosis for a Radioactive Earth*. London: Women's Press Ltd., 1985.
This book offers one of the earliest, and still one of the best, descriptions of ionizing radiation. Dr. Bertell was one of the people who linked leukemia in children to the x-ray exposure of their mothers while they were in the womb.

Caldicott, Helen. *If You Love This Planet: A Plan to Heal the Earth*. New York: W.W. Norton & Company, 2009. First published 1992.
A Canadian movie based upon Dr. Caldicott's activism, *If You Love This Planet*, won an Oscar in 1982 for Best Documentary. She hosts a radio program from Australia with the same name and has authored

several books, including *Nuclear Power Is Not the Answer* (2006).

Fernex, Solange, ed. *Chernobyl: Environmental, Health and Human Rights Implications, Vienna, Austria, 12–15 April 1996*. Permanent People's Tribunal. Geneva, Switzerland: International Peace Bureau, 1999.

Although not a particularly good translation, this covers the proceedings of a Permanent People's Tribunal held a week after the 1996 IAEA/WHO closed-door conference that established that four thousand people were killed as a result of the Chernobyl nuclear disaster. The tribunal listened to respected physicians and scientists who had worked in the Chernobyl area, who had kept clinical information and been completely ignored by United Nations organizations.

Gale, Robert Peter, and Eric Lax. *Radiation: What It Is, What You Need to Know*. New York: Alfred A. Knopf, 2013.

Dr. Gale is a hematologist (a specialist in blood) who has made a career of responding to accidents involving ionizing radioactivity. In spite of his experience, he argues for nuclear power as a viable energy source for the future.

Greene, Gayle. *The Woman Who Knew Too Much: Alice Stewart and the Secrets of Radiation*. Ann Arbor, MI: University of Michigan Press, 1999.

This is one of our all-time favourite books. It is a historically accurate blending of science and dirty politics with the fascinating life of a brilliant physician and scientist. It's a good place to go if you want to delve into the recent history of the politics of ionizing radiation.

Harding, Jim. *Canada's Deadly Secret: Saskatchewan Uranium and the Global Nuclear System.* Black Point, NS: Fernwood Publishing, 2007.
    This book combines mining, politics, and history. About half the book deals with the history of the nuclear bomb and mining in Saskatchewan, while the remainder delves into uranium mining and the nuclear industry worldwide.

Lown, Bernard, M.D. *Prescription for Survival: A Doctor's Journey to End Nuclear Madness.* San Francisco: Berrett-Koehler Publishers, 2008.
    A remarkable story of how the inventor of the electronic defibrillator received the Nobel Peace Prize for his work in nuclear disarmament.

Zoellner, Tom. *Uranium: War, Energy, and the Rock That Shaped the World.* London: Penguin Books, 2009.
    Tom Zoellner was interviewed by Jon Stewart on *The Daily Show* (see www.thedailyshow.com/watch/ thu-april-2-2009/tom-zoellner). Mr. Stewart referred to the book as "crazy, fascinating."

In writing the book, we frequently consulted the following websites for technical data and safety regulations, as well as for cross-referencing less scientific information and finding information on radiation in war and nuclear abolition:

Canadian Coalition for Nuclear Responsibility (CCNR):
　www.ccnr.org
Canadian Nuclear Association (CNA): www.cna.ca
Canadian Nuclear Safety Commission (CNSC):
　www.nuclearsafety.gc.ca
Fairewinds Energy Education: fairewinds.org
Health Canada: www.hc-sc.gc.ca
Health Physics Society (HPS): www.hps.org
International Atomic Energy Agency (IAEA): www.iaea.org
International Campaign to Abolish Nuclear Weapons (ICAN):
　www.icanw.org
International Physicians for the Prevention of Nuclear War
　(IPPNW): www.ippnw.org
Physicians for Global Survival (PGS): www.pgs.ca
Radiation Answers: www.radiationanswers.org
Radiation Effects Research Foundation (RERF):
　www.rerf.or.jp
United States Environmental Protection Agency (USEPA):
　www.epa.gov
World Nuclear Association (WNA): www.world-nuclear.org

Notes

PREFACE

1   Fox Butterfield, "Nobel Peace Prize Given to Doctors Opposed to War," *New York Times*, October 12, 1985, www.nytimes.com/1985/10/12/world/nobel-peace-prize-given-to-doctors-opposed-to-war.html.

2   Solange Fernex, ed., *Chernobyl: Environmental, Health and Human Rights Implications, Vienna, Austria, 12–15 April 1996*, Permanent People's Tribunal (Geneva, Switzerland: International Peace Bureau, 1999).

3   IAEA, "Chernobyl: Answers to Longstanding Questions," iaea.org/newscenter/focus/chernobyl/faqs.shtml.

4   James Lovelock, "Nuclear Power Is the Only Green Solution," *The Independent*, May 24, 2004, ecolo.org/media/articles/articles.in.english/love-indep-24-05-04.htm.

5   See Elizabeth Ward, "Solution to Pollution Is Dilution," *Green Risks* (blog), December 16, 2010, greenrisks.blogspot.ca/2010/12/solution-to-pollution-is-dilution.html. This is a common, and sometimes cynical, phrase used to describe an approach to environmental contaminants. While not using the phrase, the Canadian Nuclear Safety Commission (CNSC) describes the principle in its *Tritium Fact Sheet* while reassuring the public about tritium releases from nuclear power plants (see www.nuclearsafety.gc.ca/eng/readingroom/factsheets/tritium.cfm).

INTRODUCTION

1 Judy Pearsall and Bill Trumble, eds., *The Oxford English Reference Dictionary* (Oxford: Oxford University Press, 1996).

2 Frank W. Putnam, "The Atomic Bomb Casualty Commission in Retrospect," *Proceedings of the National Academy of Sciences of the United States of America*, May 12, 1998, www.pnas.org/content/95/10/5426. full.

3 Terry Macalister and Helen Carter, "Britain's Farmers Still Restricted by Chernobyl Nuclear Fallout," *The Guardian*, May 12, 2009, www.theguardian.com/environment/2009/may/12/farmers-restricted-chernobyl-disaster.

4 Chip Ward, "How the 'Peaceful Atom' Became a Serial Killer," *Mother Jones*, March 24, 2011, motherjones.com/environment/2011/03/nuclear-power-nrc-dangers.

5 Cathy Vakil and Linda Harvey, "Human Health Implications of the Nuclear Energy Industry," *Ontario College of Family Physicians*, May 2009, ocfp.on.ca/docs/committee-documents/human-health-nuclear-energy-industry.pdf?sfvrsn=3.

6 Ibid.

CHAPTER 1: HISTORICAL BACKGROUND

1 Unless otherwise cited from another source, the information in this first section of the chapter comes from Tom Zoellner, *Uranium: War, Energy, and the Rock That Shaped the World* (London: Penguin Books, 2009).

2 Gordon Edwards, "Uranium: Known Facts and Hidden Dangers," http://www.ccnr.org/salzburg.html ccnr.org.

3 Ibid.

4 Richard F. Mould, *A Century of X-rays and Radioactivity in Medicine* (Bristol and Philadelphia: Institute of Physics Publishing, 1993).

5 Ibid.

6 "The Nobel Prize in Physics 1903," http://www.nobelprize.org/nobel_prizes/physics/laureates/1903/.

7 The others were John Bardsen for the transistor in 1956 and for superconductors in 1972; Frederick Sanger for the insulin molecule in 1958 and for inventing a method to work out gene sequences in

deoxyribonucleic acid (DNA) in 1980; and Linus Pauling in 1954 for defining chemical bonds in complex substances and in 1962 for his antinuclear activism.

8   "The Nobel Prize in Chemistry 1911," http://www.nobelprize. org/nobel_prizes/chemistry/laureates/1911/.

9   Mould, *A Century of X-rays.*

10 Samuel J. Walker, *Permissible Dose* (Berkeley and Los Angeles: University of California Press, 2000).

11 "Thomas Edison," *Wikipedia*, en.wikipedia.org/wiki/Thomas_ Edison.

12 Walker, *Permissible Dose.*

13 Ibid.

14 Charles Glassmire, "Drink Radithor!" *Tales from the Nuclear Age* (blog), August 30, 2010, http://talesfromthenuclearage.wordpress. com/2010/08/30/drink-radithor/.

15 Mark Neuzil and William Kovarik, *Mass Media and Environmental Conflict: America's Green Crusade* (Thousand Oaks, CA: Sage Publications, 1996), Chapter 5.

16 R.H. Clark and J. Valentin, *The History of ICRP and the Evolution of Its Policies*, ICRP Publication 109, October 2008, http://www.icrp. org/docs/The%20History%20of%20ICRP%20and%20the%20 Evolution%20of%20its%20Policies.pdf.

17 Neuzil and Kovarik, *Mass Media.*

18 UC San Diego Moores Cancer Center, "History of Radiation Therapy," radonc.ucsd.edu/patient-info/Pages/history-radition-therapy.aspx.

19 "History of Radiation Therapy," *Wikipedia*, en.wikipedia.org/wiki/ History_of_radiation_therapy.

20 This was actually called the second congress by the *British Medical Journal*. Notes from that congress can be found in "Second International Congress of Medical Electrology and Radiology," *British Medical Journal* 2, no. 2180 (October 11, 1902): 1175–1176. However, a competing reference stated that the First International Congress of Medical Electrology and Radiology occurred in Barcelona in 1910. See A. Gonzalex-Sistal, "X-Rays: Making Waves in Medical Diagnosis for over a Century," Universitat de Barcelona, diposit.ub.edu/dspace/bitstream/2445/21264/1/X_Rays.pdf.

21 Walker, *Permissible Dose*.

22 Clark and Valentin, *The History of ICRP*.

23 Walker, *Permissible Dose*.

24 Ibid.

25 Radiation dermatitis cannot be cured, nor can radiation burns—at least not in the sense that the skin becomes entirely normal again. When the author (D.D.) first went into practice, she inherited patients who had been treated externally with radiation for cancer. Some of them spent the remainder of their lives in misery, dealing with burns that would cause recurrent breakdowns of the skin, resulting in painful cracks and poor healing. The evolution of radiation oncology has created a science that is much more patient-friendly.

26 Walker, *Permissible Dose*.

27 Thomas Schlich and Ulrilch Troehler, eds., *The Risks of Medical Innovation* (New York: Routledge, 2006).

28 Walker, *Permissible Dose*.

29 Note that as medicine becomes more accustomed to ionizing radiation that the allowable exposure numbers become smaller and smaller. In fact, the allowable exposure levels have decreased some seven times since x-rays were first used. More about millisieverts can be found in Chapter 2.

30 Clark and Valentin, *The History of the ICRP*.

31 The *reference man* is used for calculations assessing internal doses of radiation and is defined as having "the anatomical and physiologicals of an average individual," but more specifically, is "a 20–30 year old Caucasian male weighing 70 kg, 170 cm in height and living in a climate with an average temperature of 10 to 20 degrees [Celsius]. He is Western European or North American in habit and custom." See International Commission on Radiological Protection, *Report on the Task Group on Reference Man* (New York: Pergamon Press, 1975).

32 Arjun Makhijani, "The Use of Reference Man in Radiation Protection Standards and Guidance with Recommendations for Change," *Institute for Energy and Environmental Research* (blog), December 2008, blog.cleveland.com/health/2009/01/referenceman.pdf.

33 Radiation Effects Research Foundation, "Solid Cancer Risks among Atomic-Bomb Survivors," rerf.jp/radefx/late_e/cancrisk.html.

34 Sadao Asada, "Mushroom Cloud and National Psyches," in *Culture Shock and Japanese-American Relations: Historical Essays* (2007; repr., Columbia, MO: University of Missouri Press, 2011), 229.

35 From personal notes (D.D.) taken at the IPPNW Congress, Friday, August 23, 2012. The physician was Dr. Tetsuya Oda. He was in his nineties and this was his last public speech. He believed that it was extremely important for *hibakusha*, survivors of the atomic bombs, to give witness to the horrible effects of the bombs.

36 Gayle Green, "Science with a Skew: The Nuclear Power Industry after Chernobyl and Fukushima," *The Asian-Pacific Journal: Japan Focus*, January 2, 2012, japanfocus.org/-gayle-greene/3672.

37 Walker, *Permissible Dose.*

38 Ibid.

39 Ibid.

40 Clark and Valentin, *The History of ICRP.*

41 *Teratogenic effects* means that it affects the development of otherwise normal embryos and fetuses (babies in the womb). For example, alcohol is a known teratogenic substance; normal fetuses so exposed tend to develop what is known as fetal alcohol syndrome. Some may also remember thalidomide and its effects on babies born to mothers who took the drug—the shortened limbs and deformities were the result of the teratogenic effects of the drug.

42 Alberto Fassò and S. Rokni, "Operational Radiation Protection in High-Energy Physics Accelerators. Implementation of ALARA in Design and Operation of Accelerators," SLAC National Accelerator Laboratory, May 2009, slac.stanford.edu/cgiwrap/getdoc/slac-pub-13800.pdf.

43 Nuclear Information and Resource Service, "All Levels of Radiation Confirmed to Cause Cancer," Press Release, June 30, 2005, http://www.nirs.org/press/06-30-2005/1.

44 For a complete description of the agreement between WHO and the IAEA and what it entails, see Oliver Tickell, "Toxic Link: The WHO and the IAEA," *The Guardian*, May 28, 2009, theguardian.com/commentisfree/2009/may/28/who-nuclear-power-chernobyl.

45 "Nuclear Safety Watchdog Head Fired for 'Lack of Leadership':
Minister," *CBC News Canada*, January 16, 2008, cbc.ca/news/
canada/nuclear-safety-watchdog-head-fired-for-lack-of-leadership-
minister-1.748815.

46 "Japan to Probe Tepco Radiation Cover-Up Claim," *BBC News
Asia*, July 21, 2012, bbc.co.uk/news/world-asia-18936831.

47 Health Canada, "Occupational Exposure to Radiation," hc-sc.
gc.ca/hl-vs/iyh-vsv/environ/expos-eng.php.

CHAPTER 2: RADIATION SCIENCE

1 In a frustratingly circular definition, an *element* is comprised of
atoms that are all the same. Gold, iron, carbon, and silicon are all
examples of elements, as are plutonium and uranium.

2 In the scientific sense, *mass* is the quantity of matter that a body
contains, measured in terms of its resistance to acceleration by a
force. It is roughly interpreted as the weight of an object.

3 Tritium is formed naturally by the action of cosmic rays on the
upper atmosphere; it is also a common by-product of nuclear power
plants, formed from water. Tritium has one proton that identifies it
as hydrogen and two neutrons; it could also be called H-3.

4 "Neutron bomb," *Wikipedia*, en.wikipedia.org/wiki/Death-ray-
bomb.

5 Decay is actually measured in becquerels, however, to the lay person, the number of atoms of a substance is an adequate substitute.

6 *Electron capture* is a less frequent decay method used by radioactive
elements to find stability. In this case, the electrons "captured" may
be the very electrons released when some of the other potassium
atoms change into calcium.

7 "Atom," *Wikipedia*, en.wikipedia.org/wiki/Atom#History_of_
atomic_theory.

8 "Dalton's Atomic Theory," iun.edu/~cpanhd/C101webnotes/
composition/ dalton.html.

9 Recently, the measurement system has been undergoing a change
to the International System of Units (SI). The literature can be
very confusing because it contains old units, new units, and sometimes both.

10 A *joule* is a unit of energy. It is a *derived unit*, meaning that knowledge of other SI units is required to define it. One joule is the energy expended (or work done) when a force of one newton is applied through a distance of one metre. One newton, also a derived unit, is the force required to accelerate one kilogram of matter at the rate of one metre per second squared. Examples of nonderived units include metres, kilograms, and seconds.

11 In another circular definition, an *erg* (after the Greek word for "work," *ergon*) is a unit of energy or mechanical work equal to 10^{-7} joules. It is also a derived unit.

12 A *coulomb*, another derived unit, this time measuring electrical charge, is distinguished from amperes by being simply the number of charged particles, while amperes are the number of charged particles moving past a given point per unit of time. An *ampere* is the SI unit measuring electrical current.

13 The Manhattan Project was the project beginning in 1939, entered into by the United States, Canada, and the UK, to develop the atomic bomb. In 1946, Louis Slotin was demonstrating a dangerous procedure to two other men when a radioactive explosion occurred. The exact doses to which any of the men were exposed is unknown but it was thought that Slotin used his body to protect his two colleagues. Information about him and other deaths from the same era can be found at Atomic Heritage Foundation, "Accidents in the Manhattan Project," atomicheritage.org/index.php/component/content/92.html?task=view.

14 United Nations Scientific Committee on the Effects of Atomic Radiation, "Answers to Frequently Asked Questions," unscear.org/unscear/fr/index.html.

15 V. Padmanabhan, A. Sugunan, C. Brahmaputhran, K. Nandini, and K. Pavithran, "Heritable Anomalies among the Inhabitants of Regions of Normal and High Background Radiation in Kerala: Results of a Cohort Study," *International Journal of Health Services* 34, no. 34 (2004): 483–515.

16 United States Nuclear Regulatory Commission, "Fact Sheet on Biological Effects of Radiation," nrc.gov/reading-rm/doc-collections/fact-sheets/bio-effects-radiation.html.

17 Krista Mahr, "Japan: Are Kids Being Exposed to Too Much Radiation?" *Time, Science & Space*, May 3, 2011, science.time.

com/2011/05/03/japan-are-kids-being-exposed-to-too-much-radiation/.

CHAPTER 3: RADIATION AND THE HUMAN BODY

1 "Petkau Effect," *Wikipedia*, en.wikipedia.org/wiki/Petkau_effect.

2 Physicians (D.D. included) working in northern Saskatchewan have heard stories to this effect. As part of the African Uranium Alliance (the working name for a number of African human rights workers and activists) and IPPNW, we also heard similar stories from villagers from Bahi, Tanzania, in 2013 and from villagers in Mali in 2012.

3 It's not really their size that causes the damage. It's merely simpler, not entirely incorrect, and certainly easier to understand if "size" is used to represent the *linear energy transfer* (LET). *LET* is a measure of the force acting upon a charged particle travelling through matter. Alpha particles have high LET.

4 United States Environmental Protection Agency, "Beta Particles," *Radiation Protection*, www.epa.gov/rpdweb00/understand/beta.html#affecthealth.

5 There are more than twenty isotopes of polonium; Po-210 is the most stable, with a half-life of 138 days. It decays to nonradioactive but toxic lead.

6 Luke Harding and Ian Semple, "Polonium-210: The Hard-to-Detect Poison that Killed Alexander Litvinenko," *The Guardian*, November 6, 2013, www.theguardian.com/world/2013/nov/06/polonium-210-poison-alexander-litvinenko.

7 *Cytoplasm* is a general term for the gel-like liquid content of a cell, along with the free-floating enzymes and mitochondria that produce energy needed for the cell to function.

8 Permanent damage can be caused by many different substances. Any chemical that is known to cause cancer has done so directly, or indirectly, by affecting the DNA, mitochondria, and enzymes in cells.

9 Kate Ravilious, "Despite Mutations, Chernobyl Wildlife Is Thriving," *National Geographic News*, April 26, 2006, news.nationalgeographic.com/news/2006/04/0426_060426_chernobyl.html.

10 US National Research Council Technical Training Center, *Radiation Concepts Manual: Biological Effects of Radiation*, www.nrc.gov/reading-rm/basic-ref/teacher/09.pdf. With exceptions referenced, most of the technical information found in this segment and parts of the chapter, and some additional excellent diagrams, can be found in this downloadable manual.

11 Taken from United States Environmental Protection Agency, "Health Effects," www.epa.gov/rpdweb00/understand/health_effects.html#est_health_effects.

12 Hiroko Tabachi, David E. Sanger, and Keith Bradsher, "Japan Faces Potential Nuclear Disaster as Radiation Levels Rise," *The New York Times*, March 14, 2011, www.nytimes.com/2011/03/15/world/asia/15nuclear.html.

13 Government of Canada, "Radiation Protection Regulations," *Justice Laws Website*, laws-lois.justice.gc.ca/eng/regulations/SOR-2000-203/.

14 Health Physics Society, "Lead Garments (Aprons, Gloves, etc.)," hps.org/publicinformation/ate/faqs/leadgarmentsfaq.html.

15 Ganesh C. Jagetia, "Radioprotective Potential of Plants and Herbs against the Effects of Ionizing Radiation," *Journal of Clinical Biochemistry and Nutrition* (March 2007): 74–81, www.ncbi.nlm.nih.gov/pmc/articles/PMC2127223/.

16 Iodine supplementation engenders controversy. It cannot be taken regularly "just in case." The body requires very little iodine to function normally. Large amounts or prolonged use is probably unsafe. Besides the fact that some people can be very sensitive to the iodine itself, iodine used in excess can cause abdominal pain, bloody diarrhea, and hypothyroidism (US National Library of Medicine, "Iodine," *Medline Plus*, www.nlm.nih.gov/medlineplus/druginfo/natural/35.html).

17 Table contents modified by the author (F.O.) based on Society of Nuclear Medicine, "Beneficial Medical Uses of Radiation," www.molecularimagingcenter.org/index.cfm?PageID=7083>; American Dental Association, "Oral Health Topics," www.ada.org/2760.aspx>; Neil Savage, "X-ray Body Scanners Arriving at Airports," spectrum.ieee.org/biomedical/imaging/xray-body-scanners-arriving-at-airports.

18 J.F. Bottollier-Depois, Q. Chan, P. Bouisset, G. Kerala, L. Plawinski,

L. Lebaron-Jacobs, "Assessing Exposure to Cosmic Radiation during Long-Haul Flights," *Radiation Research* 153, no. 5 (May 2000): 526–532.

19 Based on "Radiation Safety: LNT Model versus the Hormesis Model," *Practicing Chiropractors' Committee on Radiology Protocols*, 2006, www.pccrp.org/docs/PCCRP%20Section%20VII.pdf.

20 *Canadian Nuclear FAQ—Dr. Jeremy Whitlock*, www.nuclearfaq.ca.

21 C.J. Martin, "The LNT Model Provides the Best Approach for Practical Implementation of Radiation Protection," *British Journal of Radiology* (January 2005): 14–16.

22 David Gorski, "Ann Coulter Says: Radiation Is *Good* for You!" *Science-Based Medicine*, March 21, 2011, www.sciencebasedmedicine.org/ann-coulter-says-radiation-is-good-for-you-2/.

23 T.D. Luckey, "Radiation Hormesis: The Good, the Bad, and the Ugly," *Dose-Response, An International Journal*, September 27, 2006, www.ncbi.nlm.nih.gov/pmc/articles/PMC2477686/.

24 Gorski, "Ann Coulter Says."

CHAPTER 4: RADIATION IN MEDICINE

1 The content from this segment has been gleaned from two sources: Charles Glassmire, "Drink Radithor!" *Tales from the Nuclear Age* (blog), August 30, 2010, talesfromthenuclearage.wordpress.com/2010/08/30/drink-radithor/; and Bill Kovarik and Mark Neuzil, *Mass Media and Environmental Conflict: America's Green Crusade* (Thousand Oaks, CA: Sage Publications, 1996).

2 "History of Radiation Therapy," *Wikipedia*, en.wikipedia.org/wiki/History_of_radiation_therapy.

3 "I Read Somewhere That Drinking Water with Radon in It Was Once Considered to Have Health Benefits. Is this True?" *Free Drinking Water*, www.freedrinkingwater.com/water_health/health2/31-08-radononce-used-considered-2have-health-benefits.htm.

4 "Radithor (ca. 1928)," *Radioactive Quack Cures Museum Directory*, February 17, 2009, orau.org/ptp/collection/quackcures/radith.htm.

5 "History of Radiation Therapy," *Wikipedia*.

6 Ron Winslow, "The Radium Water Worked Fine until His Jaw Came Off: Cancer Researcher Unearths a Bizarre Tale of Medi-

cine and Roaring '20s Society," *Wall Street Journal (1923–Current File)*, August 1, 1990.

7   Ibid.

8   Glassmire, *Tales from the Nuclear Age.*

9   "The Discovery of X-rays," *NDT Resource Center*, ndt-ed.org/ EducationResources/HighSchool/Radiography/discoveryxrays. htm.

10   The creation of an x-ray involves several steps, roughly described as follows: high-voltage electricity is put into a vacuum tube that energizes electrons; the electrons are aimed at a metal target (tungsten in hospital x-ray machines, molybdenum for mammography, others for specialized services); and the metal target gives off energy in the form of photons, a beam of which is an x-ray.

11   Gayle Greene, *The Woman Who Knew Too Much: Alice Stewart and the Secrets of Radiation* (Ann Arbor, MI: University of Michigan Press, 1999).

12   Rosalie Bertell, *No Immediate Danger: Prognosis for a Radioactive Earth* (London: Women's Press Ltd., 1985).

13   A *cyclotron* is a machine that accelerates charged particles to very high energies and then fires them at a target. The target absorbs them. Very specific new elements can be created. The cyclotron has been used in cancer therapy by directing the high-energy particles at tumours. They are frequently used to produce short-acting radioisotopes for PET scans and, more recently, technitium-99m for diagnostic imaging.

14   Health Canada, "Medical Isotopes—Frequently Asked Questions," hc-sc.gc.ca/dhp-mps/brgtherap/activit/fs-fi/isotopes-med-faq-eng. php.

15   In response to the Tc-99m shortage brought about by the Chalk River shutdown of 2009, the Canadian government invited tenders to provide alternate means of producing Tc-99m. Partnerships involving several universities were able to show that it could be produced in linear accelerators. See, "Canada Backs Three New Production Projects," *Isotopes & Radiation*, www.ans.org/pubs/ magazines/download/a_870.

16   "Contrast Agent for Radiotherapy CT (computed tomography scans)," *University College London Hospitals*, March 2011, www.

uclh.nhs.uk/PandV/PIL/Patient%20information%20leaflets/
Contrast%20agent%20for%20radiotherapy%20CT%20scans.
pdf.

17 Rita F. Redberg and Rebecca Smith-Bindman, "We Are Giving
Ourselves Cancer," *The New York Times*, January 30, 2014, www.
nytimes.com/2014/01/31/opinion/we-are-giving-ourselves-cancer.
html?_r=0.

18 Mark J. Eisenberg, Jonathan Afilalo, Patrick R. Lawler, Michal
Abrahamowicz, Hugues Richard, and Louise Pilote, "Cancer Risk
Related to Low-Dose Ionizing Radiation from Cardiac Imaging
in Patients after Acute Myocardial Infarction," *Canadian Medical
Association Journal* 183 (March 8, 2011): 430–436.

19 We searched many websites for patient information about the use
of radioisotopes and found sometimes very detailed information,
always accompanied with reassurances that the exposure was "low,"
or, as on this website, "no hazard to worry about." See, "Nuclear
Medicine Patient Issues—Diagnostic Nuclear Medicine," *Health
Physics Society*, hps.org/publicinformation/ate/q9412.html.

20 Ibid.

21 "Benefits of Radiotherapy Outweigh Small Increased Risk of
Second Cancer," *Ecancernews*, 2011, ecancermedicalscience.com/
news-insider-news.asp?itemid=1660.

CHAPTER 5: INDUSTRIAL USE OF RADIATION

1 The following three case studies come from the International
Atomic Energy Agency (IAEA), "Lessons Learned the Hard
Way," *IAEA Bulletin* 47, no. 2 (2004), www.iaea.org/Publications/
Magazines/Bulletin/Bull472/htmls/lessons_learned.html.

2 Robert Peter Gale and Eric Lax, *Radiation* (New York: Alfred A.
Knopf, 2013), 3–10.

3 "Gauging Devices," *Radiation Answers*, 2007, www.radiationanswers.
org/radiation-sources-uses/industrial-uses/gauging-devices.html.

4 United States Environmental Protection Agency (USEPA), "Moisture and Density, Nuclear Gauges Used in Road Construction,"
www.epa.gov/radtown/gauges.html.

5 Canadian Nuclear Safety Commission, *Regulatory Action—P. Machi-*

broda Engineering Ltd., September 24, 2013, www.nuclearsafety.
gc.ca/eng/acts-and-regulations/regulatoryaction/p-machibroda-
engineering.cfm.

6 USEPA, "Nuclear Logging," www.epa.gov/esd/cmb/Geophys-
icsWebsite/pages/reference/methods/Borehole_Geophysical_
Methods/Logging_Techniques_and_Tools/Nuclear_Logging.
htm.

7 Health Canada, "Safety Code 34, Radiation Protection and Safety
for Industrial X-Ray Equipment," www.hc-sc.gc.ca/ewh-semt/pubs/
radiation/safety-code_34-securite/index-eng.php.

8 R.N. Mukherjee, "Radiation: A Means of Sterilization," *International Atomic Energy Agency*, www.iaea.org/Publications/Magazines/
Bulletin/Bull176/17605882837.pdf.

9 A.G. Chmielewski and A.J. Bereka, *Trends in Radiation Sterilization
of Health Care Products* (Vienna, Austria: IAEA, 2008), 49, www-
pub.iaea.org/MTCD/publications/PDF/Pub1313_web.pdf.

10 Health Canada, "Frequently Asked Questions Regarding Food Irradiation," hc-sc.gc.ca/fn-an/securit/irridation/faq_food_irradiation_
aliment01-eng.php.

11 USEPA, "Mail Irradiation," www.epa.gov/radiation/sources/mail_
irrad.html.

12 A.A. Hammad, "Microbiological Aspects of Radiation Sterilization," *Trends in Radiation Sterilization of Health Care Products*
(Vienna, Austria: IAEA, 2008), 123, www-pub.iaea.org/MTCD/
publications/PDF/Pub1313_web.pdf. This online publication can
be downloaded for free and is an excellent resource for a full description of irradiators, their operation, safety features, and regulations
governing their use.

13 John Cotter, "Beef Industry to Ask Ottawa to Approve Irradiation
to Kill E. coli," *CityNews Toronto*, April 28, 2013, www.citynews.ca/
2013/04/28/beef-industry-to-ask-ottawa-to-approve-irradiation-
to-kill-e-coli/.

14 Food and Water Watch, "Irradiation Facts," www.foodandwaterwatch.
org/food/irradiation-facts/.

15 Health Canada, "Dioxins and Furans," www.hc-sc.gc.ca/hl-vs/
iyh-vsv/environ/dioxin-eng.php.

16 See "Smoke Detector," *Wikipedia*, en.wikipedia.org/wiki/Smoke_

detector. For the full text of the Albany, California, ordinance, see www.albanyca.org/index.aspx?page=926, and for an interesting and detailed text about smoke detection, see James A. Milke, *History of Smoke Detection: A Profile of How the Technology and Role of Smoke Detection Has Changed*, www.industry.usa.siemens.com/ topics/us/en/smoke-detection-knowledge-center/Documents/ HistoryofSmoke%20Detection.pdf.

17 See Nathan Chandler, "How Backscatter X-Ray Systems Work," *Electronics: How Stuff Works*, electronics.howstuffworks.com.

CHAPTER 6: NUCLEAR POWER PLANTS

1   The information for this section about Chernobyl is integrated from two sources: "Liquidators," *The Chernobyl Gallery*, chernobylgallery. com/chernobyl-disaster/liquidators/ (including the quotes from General Tarakanov); and World Information Service on Energy, "Chernobyl: Chronology of a Disaster," *Nuclear Monitor*, no. 724, March 11, 2011, www.nirs.org/mononline/nm724.pdf.

2   International Atomic Energy Agency, "Chernobyl's 700,000 'Liquidators' Struggle with Psychological and Social Consequences," August 2005, www.iaea.org/newscenter/features/chernobyl-15/ liquidators.shtml.

3   Stating that there are only 200,000 survivors twenty-five years later of an original 350,000 healthy young men suggests a high mortality in liquidators. Marianne Lavell, "With Thin Lead Shields," *National Geographic News*, April 26, 2011, news.nationalgeographic.com/ news/energy/2011/04/110426/chernobyl-25-years-liquidators-pictures/.

4   European Nuclear Society, "Nuclear Power Plants, World-Wide," www.euronuclear.org/info/encyclopedia/n/nuclear-power-plant-world-wide.htm.

5   Dwight D. Eisenhower Library, "Atoms for Peace," eisenhower. archives.gov/research/online_documents/atoms_for_peace.html.

6   World Nuclear Association, "Nuclear Power in the US," January 2014, world-nuclear.org/info/Country-Profiles/Countries-T-Z/ USA--Nuclear-Power/.

7   See Andrew Alden, "The Oklo Natural Nuclear Reactor," geology. about.com/od/geophysics/a/aaoklo.htm. Some scientists believe

that there is evidence of naturally occurring fissile reactions having occurred in Gabon over millions of years ago. The theory is that there would have been more uranium-235 at that time, so the higher concentration may have put atoms in close enough contact that a fissile reaction occurred.

8   *Wise Quotes*, http://wisequotes.org/nuclear-power-is-one-hell-of-a-way-to-boil-water.

9   World Nuclear Association, "World Nuclear Power Reactors & Uranium Requirements," February 1, 2014, www.world-nuclear. org/info/Facts-and-Figures/World-Nuclear-Power-Reactors-and-Uranium-Requirements/.

10   Ibid.

11   Ibid.

12   *Heavy water* is 99.75 per cent deuterium dioxide. *Deuterium* is a natural isotope of hydrogen, with a neutron in its nucleus in addition to the single proton of the far more common hydrogen. It forms 0.0158 per cent of hydrogen in nature.

13   Teach Nuclear, "The Candu Reactor," teachnuclear.ca/contents/cna_nuc_tech/candu-2/.

14   "Advanced Gas-Cooled Reactor," *Wikipedia*, en.wikipedia.org/wiki/Advanced_gas-cooled_reactor.

15   World Nuclear Association, "RBMK Reactors Appendix to Nuclear Power Reactors," June 2010, world-nuclear.org/info/Nuclear-Fuel-Cycle/Power-Reactors/Appendices/RBMK-Reactors/.

16   Thomas B. Cochran, Harold A. Feiveson, Walt Patterson, Gennadi Pshakin, M.V. Ramana, Mycle Schneider, Tatsujiro Suzuku, Frank N. Von Heppel, "It's Time to Give Up on Breeder Reactors," *A Research Report of the International Panel on Fissile Materials (IPFM)*, February 2010, fissilematerials.org/library/rr08.pdf.

17   Neil Reynolds, "With Thorium, We Could Have Safer Power," *The Globe and Mail*, September 10, 2012, theglobeandmail.com/globe-debate/with-thorium-we-could-have-safe-nuclear-power/article6255549/.

18   World Nuclear Association, "Thorium," November 16, 2013, www.world-nuclear.org/info/Current-and-Future-Generation/Thorium/.

19   Matthew L. Wald, "Uranium Substitute Is No Longer Needed but Its

Disposal May Pose Security Risk," *The New York Times*, September 23, 2012, nytimes.com/2012/09/24/us/uranium-233-disposal-proves-a-problem.html?_r=0.

20 "Uranium-233," readtiger.com/wkp/en/Uranium-233.

21 Cochran et al., "It's Time to Give Up."

22 "Fusion Power," *Wikipedia*, en.wikipedia.org/wiki/Fusion_power.

23 World Nuclear Association, "Nuclear Fusion Power," December 2013, www.world-nuclear.org/info/Current-and-Future-Generation/Nuclear-Fusion-Power/.

24 Gordon Edwards, "The Decline of Nuclear Power Worldwide," *Canadian Coalition for Nuclear Responsibility*, November 14, 2012, ccnr.org/blog_decline_of_nuclear_2012.pdf.

25 International Atomic Energy Agency, *Safety of Nuclear Power Plants: Design, Specific Safety Requirements No. SSR-2/1*, pub.iaea.org/MTCD/publications/PDF/Pub1534_web.pdf.

26 Ibid.

27 World Nuclear Association, "Cooling Power Plants," September 2013, world-nuclear.org/info/Current-and-Future-Generation/Cooling-Power-Plants/.

28 Scott Disavino, "Four US Power Reactors Shut & NYC Sweats during Heat Wave," *Reuters*, July 18, 2012, reuters.com/article/2012/07/18/utilities-usa-heatwave-idUSL2E8II52W20120718.

29 "Jellyfish Force Nuclear Plant Shutdown in Sweden," *Associated Press*, October 1, 2013, cbc.ca/news/world/jellyfish-force-nuclear-plant-shutdown-in-sweden-1.1875316.

30 "Justices Join Debate over Power Plants, Kills Fish," *Associated Press*, October 18, 2008, nbcnews.com/id/27251643/ns/us_news-environment/t/justices-join-debate-over-power-plants-fish-kills/.

31 International Atomic Energy Agency, *Safety of Nuclear Power Plants*.

32 "Cooling Tower," *Wikipedia*, en.wikipedia.org/wiki/Cooling_tower.

33 World Nuclear Association, "Cooling Power Plants," world-nuclear.org/info/Current-and-Future-Generation/Cooling-Power-Plants/.

34 See Simon Rogers, "Nuclear Power Plant Accidents: Listed and Ranked since 1952," *The Guardian*, March 18, 2011, theguardian.com/news/datablog/2011/mar/14/nuclear-power-plant-accidents-

list-rank. Partial meltdowns are included in this list: NRX Chalk River, Canada (1952); Windscale, UK (1957); SL-1 Idaho, USA (1961); Three Mile Island, USA (1979); Chernobyl, Ukraine (1986); Fukushima, Japan (2011).

35 Ibid.

36 "Nuclear and Radiation Accidents," *Wikipedia*, en.wikipedia.org/wiki/Nuclear_and_radiation_accidents.

37 Linda Harvey and Cathy Vakil, "Human Health Implications of Uranium Mining and Nuclear Power Generation," May 2009, pgs. ca/wp-content/uploads/2008/03/human-health-im_ation2009-21.pdf.

38 P. Kaatsch, C. Spix, S. Schmiedel, R. Schulze-Rath, A. Mergenthaler, and M. Blettner, *Epidemiologische Studie zu Kinderkrebs in der Umgebung von Kerkraftwerken (KiKK Studie)*, Sltzgitter: *Bundesamt fuer Strahlenschutz*, 2007, urn:nbn:de:0221-20100317939.

39 Rudi H. Nussbaum, "Childhood Leukemia and Cancers Near German Nuclear Reactors: Significance, Content, and Ramifications," *International Journal of Occupational and Environmental Health* 15, no. 3 (July/September 2009), ijoeh.com.

40 Comment made on September 29, 2010, during a hearing regarding the transport of radioactive steam generators from the inside of nuclear power plants through the Great Lakes to Sweden. The author's (D.D.) intervention was #10–H19.38, but the records for that date either did not get onto the website or have been removed. The commissioner was Dr. Patsy Thompson, the health and environmental expert on the panel. She was demonstrating the ingrained denial of the nuclear industry to recognize the effects of radioactivity on populations. The statement was live-streamed and heard by the author's (D.D.) bookkeeper, Bill Curry, in Saskatchewan. It is a travesty that she represents the Canadian voice on health safety.

41 Ian Fairlie, "New French Study on Childhood Leukemias Near Nuclear Power Plants," January 20, 2012, ianfairlie.org/uncategorized/new-french-study-on-childhood-leukemias-near-nuclear-power-plants/.

42 Carl Behrens and Mark Holt, "Nuclear Power Plants: Vulnerability to Terrorist Attack," *CRS Report for Congress*, February 4, 2005, CRS-4, www.fas.org/irp/crs/RS21131.pdf.

43 Ibid.

44 International Atomic Energy Agency, "Factsheets & FAQs, Managing Radioactive Waste," iaea.org/Publications/Factsheets/English/manradwa.html.

45 Allison Macfarlane, "It's 2050: Do You Know Where Your Nuclear Waste Is?" *Bulletin of the Atomic Scientist* 67, no. 30 (2011): 30–36.

46 "Dry Cask Storage," *Wikipedia*, en.wikipedia.org/wiki/Dry_cask_storage.

47 Macfarlane, "It's 2050," 31.

48 NWMO, "Adaptive Phased Management," nwmo.ca/faq_adaptive_phased_management.

49 George Mack, "Patrick Moore: From Greenpeace Dove to Nuclear Power Phoenix," *The Energy Report*, September 29, 2011, theenergyreport.com/pub/na/patrick-moore-from-greenpeace-dove-to-nuclear-power-phoenix.

50 George Monbiot, "The Moral Case for Nuclear Power," August 8, 2011, monbiot.com/2011/08/08/the-moral-case-for-nuclear-power/.

51 Gordon Edwards, "The Decline of Nuclear Power Worldwide," November 14, 2012, ccnr.org/blog_decline_of_nuclear_2012.pdf.

52 "Is Nuclear Power a Global Warming Solution?" *Time for Change*, timeforchange.org/pros-cons-nuclear-power-global-warming-solution.

53 See "Uranium Mining," *CESOPE*, ippnw.org/pdf/2013-conference-uranium-mining-tanzania.pdf. From the author's own notes (D.D.) taken at the conference. This site contains a summary of the conference in video form plus links to some of the conference papers.

54 World Nuclear Association, "Three Mile Island Accident," January 2012, world-nuclear.org/info/Safety-and-Security/of-Plants/Three-Mile-Island-accident/.

55 Ibid.

56 See Sue Sturgis, "Startling Revelations about Three Mile Island Disaster Raise Doubts over Nuke Safety," *Global Research*, July 24, 2011, globalresearch.ca/startling-revelations-about-three-mile-island-disaster-raise-doubts-over-nuke-safety/25757. The figures in this paragraph, including the reference to Steven Wing, are all taken from this website.

57 "Novarka: Encasing the Unsafe Chernobyl Reactor in a Huge New Arch—Video," *The Guardian*, April 19, 2011, theguardian.com/ environment/video/2011/apr/19/novarka-chernobyl-reactor-arch-video.

58 Ibid.

59 "Novarka and Chernobyl Project Management Unit Confirm Cost and Time Schedule for Chernobyl New Safe Confinement," *European Bank for Reconstruction and Development*, April 2011, ebrd. com/pages/news/press/2011/110408e.shtml.

60 David Biello, "Partial Meltdowns Led to Hydrogen Explosions at Fukushima Nuclear Power Plant," *Scientific American*, March 14, 2011, scientificamerican.com/article/partial-meltdowns-hydrogen-explosions-at-fukushima-nuclear-power-plant/. The information in the preceding paragraphs was gleaned from this and World Nuclear Association, "Fukushima Accident," January 13, 2014, world-nuclear.org/info/Safety-and-Security/Safety-of-Plants/ Fukushima-Accident/.

61 WNA, "Fukushima Accident."

62 Justin McMurry, "Fukushima Residents May Never Go Home Say Japanese Officials," *The Guardian*, November 12, 2013, theguardian. com/environment/2013/nov/12/fukushima-daiichu-residents-radiation-japan-nuclear-power.

63 The international established standard of acceptable exposure to industrial ionizing radiation is one millisievert per year for everyone who is not working with ionizing radiation; workers—in mining, power plants, refining, or labs—are permitted five millisieverts per year. See Chapter 2.

64 Shin-ichi Suzuki, "Highly Sophisticated Ultrasound Thyroid Examination Used in the Health Management Survey," paper presented at Fukushima Medical University, February 25, 2013, fmu. ac.jp/radiationhealth/conference/presentation/day1/1109.pdf.

65 The quotes and other material in this paragraph are taken from the author's (D.D.) notes from lectures and conversations with physicians in the Fukushima prefecture in August 2012. The presentation by Dr. Suzuki is practically unchanged from what was presented to our International Physicians for the Prevention of Nuclear War delegation.

66 "Disturbing Thyroid Cancer Rise in Fukushima Minors," *RT Question More*, August 21, 2013, rt.com/news/fukushima-children-thyroid-cancer-783/"Dist.

67 "Dr. John Gofman," *Whale*, whale.to/a/gofman.html.

CHAPTER 7: URANIUM MINING

1 Unless otherwise cited from another source, see Marley Shebala, "Poison in the Earth," *Navajo Times*, July 23, 2009, navajotimes.com/news/2009/0709/072309uranium.php.

2 This is an example of where the older unit, *curie*, is used instead of the SI unit, *becquerel*. This preference seems to occur fairly frequently, perhaps because the forty-six curies becomes an unwieldy 1780 gigabecquerels. As mentioned in Chapter 2, this creates difficulties for professionals and nonprofessionals alike.

3 While most of the material for this case study comes from the *Navajo Times* article cited above, for supporting material, see Laurie Wirt, "Radioactivity in the Environment; a Case Study of the Puerco and Little Colorado River Basins, Arizona and New Mexico," *USGS Water-Resource Investigations Report 94-4192*, 1994, pubs.er.usgs.gov/publication/wri944192.

4 World Nuclear Association (WNA), "Uranium Mining Overview," May 2012, world-nuclear.org/info/Nuclear-Fuel-Cycle/Mining-of-Uranium/Uranium-Mining-Overview/.

5 "Nuclear Power Plant," *Wikipedia*, en.wikipedia.org/wiki/Nuclear_power_plant.

6 The major mining companies involved in the extraction of uranium are Cameco, BHP Billiton, Areva, Rio Tinto, ARMZ Uranium Holding Co. (AtomRedMet Zoloto), and Energy Resources of Australia (ERA). These companies often partner with one another or with governments. See, WNA, "Uranium Mining Overview."

7 Cathy Vakil and Linda Harvey, "Human Health Implications of Uranium Mining and Nuclear Power Generation" May 2009, pgs. ca/wp-content/uploads/2008/03/human-health-im_ation2009-21.pdf.

8 WNA, "World Uranium Mining Production," July 2013, world-nuclear.org/info/Nuclear-Fuel-Cycle/Mining-of-Uranium/World-Uranium-Mining-Production/.

9 Cameco, "McArthur River," cameco.com/mining/mcarthur_river/.

10 WNA, "World Uranium Mining Production."

11 Ibid.

12 Ibid.

13 Ibid.

14 Gordon Edwards, "Uranium: The Deadliest Metal," *Canadian Coalition for Nuclear Responsibility*, ccnr.org/uranium_deadliest.html.

15 WNA, "Environmental Aspects of Uranium Mining," February 2011, world-nuclear.org/info/Nuclear-Fuel-Cycle/Mining-of-Uranium/Environmental-Aspects-of-Uranium-Mining/.

16 See WNA, "Nuclear Fuel Fabrication," October 2013, world-nuclear.org/info/Nuclear-Fuel-Cycle/Conversion-Enrichment-and-Fabrication/Fuel-Fabrication/.

17 WNA, "Military Warheads as a Source of Nuclear Fuel," January 2014, world-nuclear.org/info/Nuclear-Fuel-Cycle/Uranium-Resources/Military-Warheads-as-a-Source-of-Nuclear-Fuel/.

18 Dale Dewar, Linda Harvey, and Cathy Vakil, "Uranium Mining and Health," *Canadian Family Physician*, May 2013, cfp.ca/content/59/5/469.

19 W.J. Riley, A.L. Robinson, A.J. Gadgil, and W.W. Nazaroff, "Effects of Variable Wind Speed and Direction on Radon Transport from Soil into Buildings: Model Development and Exploratory Results," *Refdoc*, cat.inist.fr/?aModele=afficheN&cpsidt=1744135.

20 "Lac-Mégantic Train Disaster Voted Top News Story of 2013," *CBCnews Montreal*, December 25, 2013, cbc.ca/news/canada/montreal/lac-mégantic-train-disaster-voted-top-news-story-of-2013-1.2476212.

21 Steve Almasy, "North Dakota Train Collision Ignites Oil Cars; Fire to Burn Out," *CNN*, December 30, 2013, cnn.com/2013/12/30/us/north-dakota-train-fire/.

22 Gordon Edwards, "Fueling the Nuclear Arms Race ~ and How To Stop It," *Ploughshares Monitor* VI, no. 2 (June 1985), ccnr.org/non_prolif.html.

23 Saskatchewan Mining Association, "Uranium Cluff Lake Mine," 2012, saskmining.ca/commodity-info/Commodities/38/uranium.html.

24 Peter Diehl, "Uranium Mining in Eastern Germany: The WISMUT Legacy," *WISE Uranium Project*, April 17, 2011, wise-uranium.org/uwis.html.

25 "Environmental Problems of Northern Eurasia, The Mayak Facility in the Southern Urals," *Russian Nature*, rusnature.info/env/19_4.htm.

26 The author (D.D.) did this herself while working in Uranium City, Saskatchewan. A nurse and D.D. regularly rode their bikes out to, and all around, the Beaver Lodge Lake site. There was no signage suggesting that it might have been, and still was, radioactive.

27 Edwards, "Uranium: The Deadliest Metal."

28 Don Munson, "America's Greatest Atomic Radiation Crisis," *Colorado Indymedia*, December 1, 2007, colorado.indymedia.org/node/340.

29 Peter Prebble and Ann Coxworth, "The Government of Canada's Legacy of Contamination in Northern Saskatchewan Watersheds," *Canadian Centre for Policy Alternatives–SK, Saskatchewan Notes*, July 2013, policyalternatives.ca/sites/default/files/uploads/publications/ Saskatchewan%20Office/2013/07/SKnotes_Govt_Legacy_Contamination_Watersheds.pdf.

30 Ibid.

31 "Report from Church Rock: A Uranium Legacy Update," *La Jicarita*, September 10, 2012, http://lajicarita.wordpress.com/2012/09/10/report-from-church-rock-a-uranium-legacy-update/.

32 Ibid.

CHAPTER 8: TRANSPORTATION OF RADIOACTIVE MATERIALS

1 "Emergency Response Guidebook 2012," *Radioactive Materials*, www.apps.tc.gc.ca/saf-sec-sur/3/erg-gmu/erg/guidepage.aspx/guide163.

2 In fact, it is highly unlikely that one of these barrels would explode; however, experience tells us that even the most unlikely events can, and do, occur. The presence of the truck carrying ammonia is probably a bigger concern. If it were following hazardous materials regulations, it would not be on the road at that time.

3   Ian Jones (retired Manitoba PA, FF/EMT-P), personal communication with the author (D.D.), January 20, 2013.

4   U. Stahmer, *Transport of Used Nuclear Fuel—A Summary of Canadian and International Experience*, Nuclear Waste Management Organization, April 2009, 5, www.nwmo.ca.

5   Ibid.

6   Canadian Nuclear Safety Commission,"Regulating the Packaging and Transport of Nuclear Substances in Canada," www.cnsc.ca.

7   Nuclear Waste Management Organization, "Safe and Secure Transportation of Canada's Used Nuclear Fuel," www.nwmo.ca.

8   Health Physics Society, "Instrumentation and Measurements," hps.org.

9   International Atomic Energy Commission, "Threat Assessment & Risk-Informed Approach for Implementation of Nuclear Security Measures for Nuclear and Other Radioactive Material Out of Regulatory Control," www-ns.iaea.org.

10 CNSC, "Regulating the Packaging and Transport of Nuclear Substances in Canada," nuclearsafety.gc.ca/eng/readingroom/factsheets/packaging-and-transport-of-nuclear-substances.cfm.

11 "Transportation of Radioactive Materials," *Alberta Transportation*, 2009. Alberta Transport states that its regulations are "in accordance with those of the IAEA." They describe the responsibilities of both the consignor and the operator (carrier) of the vehicle.

12 John Spears, "Trucks with Radioactive Cargo Fail Inspections," *The Star*, November 15, 2013, www.thestar.com.

13 Canadian Environmental Assessment Agency, "Undertaking #61—Ministry of Transportation," December 2013, www.ceaa-acee.gov.ca.

14 CNSC, "Regulating the Packaging and Transport."

15 John MacLeod, "Bomb-Grade Uranium To Be Shipped Secretly from Chalk River, Ontario Nuclear Plant to US," *National Post*, February 11, 2013, news.nationalpost.com.

16 Ibid.

17 Nuclear Waste Management Organization, "Ensuring Safe Transportation of Used Nuclear Fuel," www.nwmo.ca/uploads_managed/MediaFiles/1959_backgrounder_ensuringsafetransportation2012.pdf.

18 "United States Atomic Energy Commission," *Wikipedia*, en. wikipedia. org/wiki/United_States_Atomic_Energy_Commission.

19 State of Nevada, "Reported Incidents Involving Spent Nuclear Fuel Shipments 1949 to Present," www.state.nv.us/nucwaste/trans/ nucinco1.htm.

20 Pierre Sadik, "Nuclear Waste Transportation Accidents in the US," Fact Sheet, www.nuclearactive.org/graphix/transport_accidents. pdf.

21 Marvin Resnikoff, *The Next Nuclear Gamble: Transportation and Storage of Nuclear Waste* (New York: Council on Economic Priorities, 1983).

22 Green Party of Canada, "CEAA Cuts Provoke Outrage," www. greenparty.ca/media-release/2011-07-21/ceaa-cuts-provoke-outrage.

23 Personal project of the author (D.D.) and personal communications with the respective offices, October 2012.

24 Both the NWMO and CNSC advance this claim (see www.nwmo. ca or www.iaea.org).

CHAPTER 9: RADIATION IN WAR

1 Edward Wilson, "Thank You Vasili Arkhipov, the Man Who Stopped Nuclear War," *The Guardian*, October 27, 2012, theguardian.com/ commentisfree/2012/oct/27/vasili-arkhipov-stopped-nuclear-war.

2 Chris Hubbuch, "False Alarm: How a Bear Nearly Started a Nuclear War," *La Crosse Tribune*, January 30, 2009, lacrossetribune.com/ news/article_bc6f4da6-a89c-5d7d-bf0a-e41150753b62.html.

3 "The Actions of Stanislav Petrov Prevented a Worldwide Nuclear War," *Bright Star Sound*, brightstarsound.com.

4 Josh White, "In Error, B-52 Flew over US with Nuclear-Armed Missiles," *The Washington Post*, September 6, 2007, washingtonpost. com/wp-dyn/content/article/2007/09/05/AR2007090500762.html.

5 World Nuclear Association, "Nuclear-Powered Ships," January 2014, world-nuclear.org/info/Non-Power-Nuclear-Applications/ Transport/Nuclear-Powered-Ships/.

6 Chunyan Ma and Frank von Hippel, "Ending the Production of Highly Enriched Uranium for Naval Reactors," *The Nonprolifera-*

tion Review, Spring 2001, ns.miis.edu/npr/pdfs/81mahip.pdf.

7   Thomas Milsen, "Urgent to Lift Dumped K-27 Nuclear Sub," *Barents Observer*, September 25, 2012, barentsobserver.com/en/nature/urgent-lift-dumped-k-27-nuclear-sub-25-09.

8   "Disasters: Nuclear Accidents," *Pollution Issues*, pollutionissues.com/Co-Ea/Disasters-Nuclear-Accidents.html.

9   George Montgomery, "The *Komsomolets* Disaster," *Studies in Intelligence*, 1995, www.fas.org/man/dod-101/sys/ship/row/rus/si-montgomery.htm.

10 Steinar Hoibraten, Per E. Thoresen, and Are Haugan, "The Sunken Nuclear Submarine *Komsomolets* and Its Effects on the Environment," *The Science of the Total Environment* 202 (1997): 67–78.

11 Congress of the United States, Office of Technology Assessment, *Nuclear Wastes in the Arctic: An Analysis of Arctic and Other Regional Impacts from Soviet Nuclear Contamination*, OTA-ENV-623 (Washington, DC: US Government Printing Office, September 1995), ota.fas.org/reports/9504.pdf.

12 Ibid.

13 "1993—Dumping of Radioactive Waste at Sea Gets Banned," *Greenpeace International*, September 13, 2011, greenpeace.org/international/en/about/history/Victories-timeline/radioactive-dumping/.

14 United States Environmental Protection Agency, "London Convention," water.epa.gov/type/oceb/oceandumping/dredgedmaterial/londonconvention.cfm.

15 "Cold War Nuclear Legacy Dismantlement in Sayda Bay—Mission: Possible," *Bellona*, August 12, 2013, bellona.org/news/uncategorized/2013-08-cold-war-nuclear-legacy-dismantlement-in-sayda-bay-mission-possible. Some excellent photographs of the Sayda site and the work that occurs there are found at "Status of Construction on the Long-Term Storage Facility of Reactor Compartments in the Sayda-Bay," *EWN, 23rd Meeting of the CEG IAEA*, Rome, Italy, October 7–9, 2009, iaea.org/OurWork/ST/NE/NEFW/CEG/documents/CEG_23/4-6%20Mietann%20Eng.pdf.

16 "Movement on Dismantlement of Decrepit *Lepse* Nuclear Service Ship Finally on Horizon," *Bellona*, September 12, 2013, bellona.org/news/arctic/russian-nuclear-icebreakers-fleet/2013-12-movement-dismantlement-decrepit-lepse-nuclear-service-ship-finally-horizon.

17 Frans Berkhout and William Walker, "Transparency and Fissile Materials," in *Fissile Materials: Scope, Stocks and Verification*, Disarmament Forum, Published by Federation of American Scientists, 73–84, fas.org/nuke/control/fmct/2e-berkh.pdf.

18 "Depleted Uranium," *Wikipedia*, en.wikipedia.org/wiki/Depleted_uranium.

19 The US Veteran Affairs office prefers to use the term "chronic multisystem illness" or "undiagnosed illnesses" to refer to the spectrum of symptoms otherwise called Gulf War Syndrome. In 1992, returned soldiers started complaining of fatiguability, aching joints, headaches, weight loss, irritability, rashes, and hair loss. Additionally, there was an increase in stillbirths and spontaneous abortions among them. See "Gulf War Syndrome," *Health Central*, www.healthcentral.com/encyclopedia/408/320.html.

20 Jason Grotto and Tim Jones, "Agent Orange: Part 1 of 5 Agent Orange's Lethal Legacy: For US, a Record of Neglect," *Chicago Tribune News*, December 4, 2009, articles.chicagotribune.com/2009-12-04/health/chi-agent-orange1-dec04_1_defoliants-u-s-veterans-vietnam-veterans.

21 Observers from the IAEA know when a country is trying to create a plutonium nuclear bomb because it removes its reactor fuel before the uranium is "spent." It must prevent a buildup of Pu-240 while harvesting Pu-239. The Pu-240 prevents Pu-239 from sustaining a fissile reaction. See "Nuclear Weapons Proliferation," *NuclearInfo.net*, nuclearinfo.net/Nuclearpower/WebHomeNuclearWeaponsProliferation.

22 Ibid.

23 International Commission on Nuclear Non-proliferation and Disarmament, *Eliminating Nuclear Threats: A Practical Agenda for Global Policymakers* (Canberra, Australia: Paragon, 2009), box 4-1.

24 Ibid.

25 Zachary Keck, "Why Countries Build Nuclear Weapons in the 21st Century," *The Diplomat*, July 3, 2013, http://thediplomat.com/2013/07/why-countries-build-nuclear-weapons-in-the-21st-century/.

26 "What Nuclear Weapons Cost Us," *Ploughshares Fund*, ploughshares.org/what-nuclear-weapons-cost-us.

27 All of the numbers in Table 9.1 are approximate estimates. Of all of these weapons, 4,300 are considered operational, ready to fire, and about 1,800 US and Russian warheads are on high alert, ready for

use within less than thirty minutes. For more details on the types of warheads and details about each country, see "Status of World Nuclear Forces," *Federation of American Scientists*, www.fas.org/programs/ssp/nukes/nuclearweapons/nukestatus.html.

28 "A Photo-Essay on the Bombing of Hiroshima and Nagasaki," *Modern American Poetry*, www.english.illinois.edu/maps/poets/g_l/levine/bombing.htm.

29 Big Ivan was actually designed to yield one hundred megatonnes. Concern over the fallout in populated areas resulted in a last-minute order by Premier Khrushchev to decrease the yield. See "*Big Ivan*, The Tsar Bomba ('King of Bombs')," *The Nuclear Weapon Archive*, September 3, 2007, nuclearweaponarchive.org/Russia/TsarBomba.html.

30 Andrea McKillop, "Missing Presumed Dangers—the World's Missing Nuclear Materials," *Nuclear Pledge*, nuclearpledge.com/newest1.html.

31 Unless otherwise cited, this section relies on Alan F. Phillips, "The Effects of a Nuclear Bomb Explosion on the Inhabitants of a City," Pamphlet published by Physicians for Global Survival, Canada, June 2003, pgs.ca/wp-content/uploads/2007/11/1bomb.pdf.

32 Christine C. Harwell, ed., "Experiences and Extrapolations from Hiroshima and Nagasaki," chap. 6 in *Environmental Consequences of Nuclear War Volume II: Ecological and Agricultural Effects*, ed. M.A. Harwell and T.C. Hutchinson (New York: John Wiley & Sons, 1985), 427–467, dge.stanford.edu/SCOPE/SCOPE_28_2/SCOPE_28-2_3.2_Chapter6_427-467.pdf.

33 Unless otherwise cited, this section relies on Steven Starr, "Deadly Climate Change from Nuclear War: A Threat to Human Existence," *Nuclear Files*, nuclearfiles.org/menu/key-issues/nuclear-weapons/issues/effects/PDFs/starr_climate_change.pdf. For more on the effects of nuclear weapons blasts and nuclear war, see also "Catastrophic Humanitarian Consequences of Nuclear Weapons and Nuclear War," *IPPNW*, www.youtube.com/watch?v=UXZIxeVnlyl&feature-youtube.

34 Palash Ghosh, "India-Pakistan Nuclear War Would Kill 2 Billion People, End Civilization: Report," *International Business Times*, February 10, 2014, www.ibtimes.com/india-pakistan-nuclear-war-would-kill-2-billion-people-end-civilization-report-1503604.

35 Jeffrey Masters, "The Effect of Nuclear War on Climate," wunderground.com/resources/climate/nuke.asp.

36 "War and International Humanitarian Law," *ICRC War and Law*, October 29, 2010, icrc.org/eng/war-and-law/.

37 Ibid.

38 "Working Towards the Elimination of Nuclear Weapons," *Council of Delegates 2011: Resolution 1, ICRC*, November 26, 2011, icrc.org/eng/resources/documents/resolution/council-delegates-resolution-1-2011.htm.

39 George Bunn, "The Nuclear Nonproliferation Treaty: History and Current Problems," *Arms Control Association*, December 2003, armscontrol.org/act/2003_12/Bunn.

40 David Alexander, "US Nuclear Weapon Plans to Cost $355 Billion over a Decade: CBO Report," *Reuters*, December 20, 2013, reuters.com/article/2013/12/20/us-usa-nuclear-arms-idUSBRE9 BJ1FH20131220.

41 The five permanent members are China, France, the Russian Federation, the United Kingdom, and the United States.

42 These NGOs include the Albert Schweitzer Institute, Global Security Institute, International Association of Lawyers against Nuclear Arms, International Network of Engineers and Scientists for Global Responsibility, International Peace Bureau, International Physicians for the Prevention of Nuclear War, Nuclear Age Peace Foundation, and Woman's International League for Peace and Freedom. See, *Middle Powers Initiative*, middlepowers.org.

43 "Middle Powers Initiative," *Wikipedia*, en.wikipedia.org/wiki/Middle_Powers_Initiative.

44 "Ban History," *International Campaign to Ban Landmines*, icbl.org/index.php/icbl/Treaty/MBT/Ban-History.

45 "A Nuclear Weapons Convention: The Time Is Now," Hiroshima 2020 Conference, *Canadians for a Nuclear Weapons Convention*, nuclearweaponsconvention.ca/hiroshima-2020-conference.

46 Ernie Regehr, "Canada's Parliament Endorses a Nuclear Weapons Convention," *Disarming Conflict*, December 10, 2012, disarmingconflict.ca/2010/12/10/canada's-parliament-endorses-a-nuclear-weapons-convention/.

CHAPTER 10: CONCLUSION

1 "Chernobyl 'Caused Sweden Cancers,'" *BBC News*, November 20, 2004, news.bbc.co.uk/2/hi/europe/4028729.stm.

2 Indian delegation to IPPNW conference, "Aiming for Prevention: International Medical Conference on Small Arms, Gun Violence, and Injury," personal communication with author (D.D.), September 28 and 29, 2001. We don't have the names of the three men, but they were anxious about what they had discovered, which they also felt was being kept secret.

3 "South Korea Issues Reassurance over Radiation Levels in Its Seafood," *Undercurrent News*, August 5, 2013, undercurrentnews.com/2013/08/05/south-korea-issues-reassurance-over-radiation-levels-in-its-seafood/; Phil Stewart, "Pentagon Maps Japan Radiation, Says U.S. Personnel Safe," *Reuters*, September 5, 2012, reuters.com/article/2012/09/05/us-usa-japan-radiation-idUSBRE8841JV20120905.

Index